SALVAGING OLD BARNS & HOUSES

Tear It Down & Save the Pieces

Lawrence & Kathleen Abrams

STERLING PUBLISHING CO., INC.   NEW YORK

Distributed in the U.K. by Blandford Press

Dedication

We dedicate this book to our agent, who salvaged the idea for us.

Acknowledgments

While we were researching for this book, many workers and homeowners shared their time and experiences with us. We are grateful to everyone who offered suggestions, and wish we could mention all those people here. Especially, we would like to thank the following people: Bud and June Agger; Don Arndt; Harold and Hildegarde Engel; William and Jan Gabler; Everett Kneiss; James Krueger; Leon Lemma; Harry Penrose; Stephen and Barbara Portch; Gene Roetz; and Ralph Ryan.

Library of Congress Cataloging in Publication Data

Abrams, Lawrence F.
  Salvaging old barns and houses.

  Includes index.
  1. Wrecking.   2. Barns—Salvaging.   3. Buildings—
Salvaging.   4. Building materials—Recycling.
I. Abrams, Kathleen S.   II. Title.
TH153.A24   1983   690'.8   82-19330
ISBN 0-8069-7666-7 (pbk.)

Contents

Introduction

Our interest in salvaged materials began innocently enough. We purchased a picture frame made from salvaged barn wood, framed a watercolor in it, and hung the painting in our living room. The more we looked at the frame, the more we appreciated the warm tones of the weathered wood. We admired its deep grooves and found ourselves talking respectfully about its age and endurance. We wanted to work more salvaged materials into the decoration of our home.

We tracked down a hand-hewn barn beam for a fireplace mantel, but even that did not satisfy us. Soon we had paid for the salvage rights on a horse barn and were committed to a full-fledged salvage project.

We did not know anything about tearing down a building. We chose the barn because: (1) it was picturesque (whatever that had to do with a salvage project); (2) the wood had weathered beautifully (that had a bit more application to our purpose); and (3) the weather vane and lightning rod on the roof were still intact, though they were a very small part of the whole salvage project.

We still wonder sometimes how we survived that first salvage project, but we certainly learned a lot. We learned what to look for in a building to be salvaged (its picturesque qualities are not a high priority). We learned how to hire workers and what safety precautions to take. As the summer wore on, we developed a system for dismantling a structure and salvaging materials. Through experimentation, we learned the best ways to transport and store salvage and dispose of debris. Although we had originally thought only of ripping the boards off the barn and decorating our home, we learned that a salvage project is complicated by numerous factors.

As we shared our experiences of the "summer of the barn," we learned that many people had salvaged buildings. Most of them had approached their projects in the same way we had approached our barn—in blissful ignorance of the magnitude of the undertaking. The problems and frustrations they recalled sounded like a replay of the ones we had encountered.

Salvaging was evidently a popular pastime, but it seemed that few people had any training in the work before they began the projects. We decided to share our knowledge of salvaging, with the hope that many readers would be spared some of the problems we encountered with our first salvage project.

While researching this book, we talked to many experienced salvagers. We observed them on salvage sites and worked beside many of them. We met people who salvage buildings for their own use, and those who prefer to sell the materials. We talked to hobbyists and to people whose business is salvage. We learned many things from these people which, combined with our own experiences, are expressed in this book. But one thing we learned above all others: No two buildings are exactly alike, and no two salvagers work in exactly the same way. One worker, for example, may value speed. He salvages the best materials and topples the rest into a debris pile. A few damaged boards in the process and rubble on the site are not important to him. Another worker may save all materials. He will take the time to pound nails out of boards and straighten them for later use. He stacks his salvage in neat piles and removes debris regularly.

The approach to dismantling the structure also varies. One worker may methodically dismantle an entire building from the top down, while another may collapse the structure to work at ground level. Even the tools used vary. One worker's primary tools may be the nail puller and the pry bar, while another may choose the maul.

In this book, we discuss as many different structures and techniques as possible. We present safety techniques, and attempt to anticipate problems you may encounter in starting and working on a salvage project. Remember, this book is intended as a guide; we cannot anticipate every situation a salvager may encounter. We hope the advice given can be applied (perhaps with variation) to your own salvage projects.

You may notice that some of the salvagers photographed do not wear the hat, gloves, or other items of dress we recommend in the chapter on safety. We photographed our salvagers as they worked. In each case, weigh the trade-offs, and decide whether to wear protective clothing for greater safety or enjoy the freedom of movement that bare hands and short-sleeved shirts provide.

Finally, we have included some ideas on using salvaged materials in room design. If, like us, you think beyond the dust, dirt, and physical labor of a salvage project to the appeal of the products used in decorating your house, your salvage project will probably be easier. In the last section of the book, we have included ideas for using beams as fireplace mantels and ceiling rafters, boards for panelling, and miscellaneous salvage in room decoration. Although we have tried to capture the designs in black-and-white photographs, we realize that the warmth and interest of salvage materials is best experienced firsthand.

As we talked to people about their room designs, we were most impressed by the originality of the designs. Although the materials were often similar, each person had incorporated them uniquely into his own home. Using your imagination for decorating with the materials you salvage is part of the fun of a salvage project.

Whatever your reasons for undertaking a salvage project, we wish you a safe and efficient venture, and hope this book offers insight and inspiration for your dismantling and remodelling projects.

Starting a Salvaging Project

Consider Your Reasons for Undertaking This Project

Before beginning a salvage project, consider the reasons you have for wanting to salvage the building. Although salvaging sounds easy, it is actually a very time-consuming, complicated process. After going through all the work of salvaging a building, you will want to have the materials you need from the salvage when you finish with the project.

There are several reasons for salvaging an old building. One reason is that the building is in the way. The materials in the building are good and you don't want to destroy them, but the structure itself is not useful to you. Perhaps you have bought a city lot with a small house inadequate for your needs. You want to remove the building and build another structure on the site. If you take the structure apart piece by piece, you can sell the salvaged doors, windows, bricks, and lumber to help offset the cost of your new building project.

You may also want to remove structures when purchasing rural property. This kind of property often has numerous old sheds and barns in various stages of disrepair. You may consider the structures an eyesore and a hazard. Realizing the value of the material present in the structures, you may decide the best alternative is to remove the valuable boards, beams, and other popular items and sell the materials. In either case, your primary concern is removing the structure from your land.

The second reason you may decide to tear down a building is that you need the salvage as an inexpensive source of building materials. Dilapidated sheds and outdated houses can become modern garages and additions (like recreation rooms or extra bedrooms) to existing houses. This kind of salvage can be used to improve an apartment building you own, or to build a hunting cabin. The cost of new building materials often makes a salvage project a real bargain.

Probably the most popular reason for salvaging an old structure is that the materials used to build it are unique. Weathered barn wood and hand-hewn beams are very popular today as decorating materials for modern homes. Brass light fixtures, often a standard in even the most inexpensive buildings at the turn of the century, are valuable salvage items today. Many people also prefer old brick to new brick, and find the hardwood flooring and wooden banisters they can salvage from an older structure more solid than those they can purchase today. And etched- or leaded-glass windows (again, often a standard in turn-of-the-century buildings) are today's decorating treasures.

Determining the Kind of Building You Need

Your reasons for wanting to salvage a building will help you find the best building for your purposes. Of course, if the building is on your land and in your way, you won't have to think much about the type of building it is. Your only concern is removing it, and perhaps benefitting financially from the project. If, however, you want to use the material, you should think about the kind of structure that is best for you.

Do you want to use the salvage as inexpensive building material for a garage, a hunting shack, or an addition to your home? If so, you have to find an unpretentious structure with solid basic construction. You will often be able to obtain the salvage rights to these structures free since you will be doing the owner a favor by tearing the building down for him.

Most people, on the other hand, realize the value of the barn wood and decorative features standard in once elegant older homes. You will usually pay more for the salvage rights to these buildings, so search out these kinds of structures only when you need the salvage for decorative purposes. Make sure, if your goal is decoration, that the building really has everything you need. Don't agree to tear down a whole house because you like the etched glass in the front door. Look beyond that, for other things you can use—such as hardwood flooring, a marble mantelpiece, old bricks, or ornate ceiling lights.

When you have decided what you hope to gain from salvaging a building, you will know what kind of building to look for. Old sheds, neighborhood grocery stores, and unpretentious single-family dwellings provide inexpensive building materials for garages, additions, and similar projects. Barns and other farm outbuildings are, of course, the best source of weathered wood and impressive beams. Old churches provide ornate windows, decorating banisters and mouldings, and solid hardwood floors. These structures are often constructed of brick veneer, so you will be able to salvage a lot of old brick from them.

An elegant older home usually provides the most unusual salvage. Etched-glass doors, scroll work around the eaves and porches, light fixtures, heavy wooden doors and mouldings, decorative doorknobs, and plank hardwood flooring or parquet floors are only a few of the items you can salvage from this type of structure.

Finding the Structure

When you know what kind of building you need, your next step is to find one you can salvage. You may be lucky and know the owner of a structure that fits your needs, who coincidentally also wants the building torn down. But more likely, you will have to search for one.

Newspaper ads are the most helpful way to search for a structure. First, consider the location where you are likely to find the type of building you need. Next, place your ad in a newspaper that reaches that locale. If you need weathered wood, a newspaper or shopper with a rural circulation should be your choice. If you need the salvage from a large old home, you will probably have more success in a small town or an urban area where these houses have been allowed to deteriorate.

Word the ad so it describes the type of building you want and the stipulations by which you will tear it down. Here are some examples:

Will tear down small shed or house for salvaged materials. Reliable worker. Phone 724-8312 after 3:30 p.m.	Will tear down small barn, shed, or other farm building for the weathered wood. Phone 735-8642 after 5 p.m.

Another way of finding a salvageable building is to follow leads from people who know you are looking for such a structure. It always seems that all we have to do is mention to a few people that we are looking for a salvageable building, and we quickly have several leads.

Some people who might be particularly helpful to you in finding a building are realtors, contractors, building inspectors, workers in the Department of Natural Resources, and home economic or agricultural agents from the local university extension office who travel throughout rural areas a great deal. Although we have contacted a number of these people when we were searching for salvageable structures, we have received some of our best leads at places like church picnics, family reunions, and office parties.

A third way to find a structure to salvage is to locate the building on your own. A drive through a rural area, for example, might reveal many dilapidated barns and abandoned homesteads waiting to be salvaged. Take a different route home from work occasionally. You might find an old store, church, or small house that has been abandoned.

Finding a building this way presents another problem, however. You must find the owner of the structure. Sometimes neighbors can be helpful, especially if they consider the building an eyesore and want it taken down. If the building is in a rural area, a plat book (a book of township maps showing who owns each parcel of land) will be helpful. You can find plat books at the county courthouse or the local library. When you know the owner's name, you should be able to get his address from the county treasurer's office. If the building is located in town, you can call City Hall. You can usually obtain the mailing address of the owner from the city assessor's office, located in City Hall.

Questions You Should Ask the Owner of the Property

When you talk to the owner of the property, question him thoroughly about the building. Questions you might ask are:

1. Does he hold title to the building jointly with another party? (Make sure he has the right to authorize you to tear down the structure.)

2. Where is the building located? (Find out how far away the building is from your home, its proximity to a dumping site, and whether it is inside or outside the city limits.)

3. Why does the owner want the structure to come down? (Is he simply eager to get rid of the structure, or does he want to keep the materials salvaged?)

4. What is the general condition of the building? (This will give you an idea of how difficult it will be to raze, and also the worth of the materials to be salvaged.)

5. Does the owner expect you to remove the foundation as well as the frame structure? (Will you be required to landscape the area after the materials have been removed?)

6. Under what conditions does the owner expect you to tear down the building? (Does he expect you to pay him for the salvage? Will he pay you to take the building down? Can you arrange to take the building down for no charge either to you or the owner? Does he expect a portion of the salvage?)

Assessing the Structure

If the building sounds like one you could successfully salvage, you should examine the structure firsthand.

Begin with the foundation. Look for sections of stone or block that have become loosened or have shifted so much they are ready to collapse. Poured concrete walls may have buckled or cracked. If the foundation has deteriorated badly, the building may settle or shift at any time. This calls for extreme caution, and probably a modified approach.

With barns and other open-wall structures, you can easily check on the condition of the corner posts, studding, and beams. If these are correctly aligned, and the points where they join are not rotted, the basic structure should still be sound. Rotted or tilted portions should be viewed with suspicion, since the strength is no longer there.

With houses or other buildings, plaster usually conceals the wood joints. Unless water has leaked in, though, the joints should still be tight. Plaster readily shows water marks, so a quick examination will indicate the extent of the damage. Where large areas of plaster have sagged inward, punch a hole to see how much damage may be hidden inside.

The condition of the roof is crucial to the approach. Inspect it from the inside. Look for areas where water has leaked through. You may find rotted roof boards and rafters that would never support your weight. The last thing anyone needs is to fall through a roof. As you did with plastered walls, break holes through any sagging ceiling plaster so you can see the extent of the damage. By striking the roof boards and rafters with the end of a wrecking bar, you can determine how much strength remains.

9

Another reason to inspect the attic is to check for fire damage. It is truly amazing how many fires start near the chimney, often damaging rafters and roofing without the owner's knowledge. If you plan on salvaging a certain amount of board feet of roofing and rafters, you could be disappointed at the loss from fire damage. The only way to be sure is to inspect the attic first rather than asking the owner, who may be unaware of the damage.

Sometimes water leaks may have been severe enough to damage flooring or even floor joists. Tear up any carpeting or floor covering where water damage is apparent. Then you can see whether joists or flooring have been damaged by rot. The extent of the damage may be important from a safety standpoint, as well as a help in determining the monetary exchange in the contract.

Nearly all parts of the electrical, plumbing, and heating systems can be reused directly or sold as salvage items, so examine all three if these structures are to be included in the contract. Check for cracks in sinks, toilets, or bathtubs. If these are not cracked they can easily be reinstalled, but any cracks render them useless in most cases.

As a last consideration, look for obstacles such as trees, large fences, streams, or shrubs that may interfere with your work. Try to obtain written permission in the contract to remove obstructions. Shrubs immediately around the building may be included as part of the contract. If so, soak them for several days and then dig them out. Replant them as soon as you can and they should survive easily. With the current prices of mature shrubs, it would be a shame to allow them to be destroyed by piles of boards, or by stepping on them if they are in the way.

How Much Will the Materials Cost?

When you have found a building that interests you, and know the conditions by which the owner is agreeable to your tearing it down, you should figure out how much the salvage project will cost you. Often, especially in the case of old barns, the owner may expect you to pay for the salvage rights. Although his price may seem reasonable for the amount of materials you can salvage, remember the salvage rights will not be your only expense. You may have to hire other workers to help, and will probably need some kind of liability insurance. You may be required to pay for permits required by town ordinances before you begin the salvaging project. Unless you live next door or camp on the building's lawn, you will have to drive to the site. You may also find yourself making many trips with rubble to a dumping site. Usually, you will be required to pay to dump at a landfill. These dumping charges, along with the cost of driving to the site, should be figured into the cost of the salvaged materials. If you also must pay for the salvage rights, you may find your costs mounting and the materials you salvage too expensive.

Securing the Rights to Salvage the Structure

When you have found the right building and reached a verbal agreement with the owner, you should put your agreement in writing. Even if the owner is your best friend—or perhaps especially if the owner is your best friend—you should spell out the conditions surrounding the salvage project in a written statement. You may decide that a letter from you to the owner stating the exact conditions by which you will salvage the structure will suffice, but more often you will probably elect to have an attorney help you draft the agreement.

Whichever you choose, your agreement should establish and record the following:

1. Statement of ownership of the structure.
2. Amount of cash payment by you to owner for salvage rights.
3. Cash payment by owner to you for work involved.
4. Exact extent of the salvage project. (Will you be responsible for the frame structure above ground, the foundation, etc.?)
5. Who is responsible for the landscaping after the salvage is completed.
6. Who will carry the necessary liability insurance, and who will pay any fees required for permits to salvage.
7. Who owns the salvage from the structure.
8. Time limit for the salvaging project.
9. Permission from the owner to burn rubble on his property, if the city or township allows this.

This agreement should be signed by all parties involved. For example, if the property is owned by a husband and wife, both of them should sign the agreement. It should also be witnessed, and notarized.

Obtaining Permits to Begin Work

Before you begin wrecking the building, you should obtain the necessary permits. If the building is within the city limits, you can usually get the permits you need from the office of the building inspector or the zoning administrator. You may need to purchase a wrecking permit. If the building is close to the street or sidewalk, you may also need a street-privilege permit. Before a wrecking permit is issued, you will usually be required to show proof that all the utilities—gas, water, sewer, electricity, etc.— have been disconnected.

You can usually pick up the proper forms from the office of the city building inspector, and have them signed by each utility company as that service is disconnected from your building. If your building is in a rural area, contact the township's building inspector or the county courthouse to determine what wrecking permits you will need in that area.

With the proper permits and written agreements in hand, you are ready to begin the actual salvage project.

Hiring Someone to Help

Many people who salvage an old building do so by themselves. If you are taking down a shed or a small house and you have unlimited time to raze the structure, working alone may be the most enjoyable and convenient way to demolish the building.

Many demolition projects, however, are too large to handle on your own. You may also find that you cannot finish the project within the time allowed you unless someone helps you. Although occasionally friends or relatives may volunteer to help, you may also need to hire someone.

Child-Labor Laws

If you do decide to hire someone to help, be aware of the state and federal laws which regulate employment.

Before hiring anyone, become familiar with the Fair Labor Standards Act (also known as the Federal Wage and Hour Law). You can get an easy-to-read reference guide to this act from the office of the U.S. Labor Department—Wage and Hour Division, in your community. This office will be listed in your phone book under U.S. Labor Department. Your local employment service should also be able to help you secure this guide.

The Fair Labor Standards Act is a federal act that was first established in 1938, and which has been amended many times since. It provides explicit guidelines by which an employer can hire workers. If you have any questions after reading the guide, call your local Wage and Hour office. They will be happy to answer your questions.

The first thing you should familiarize yourself with is the child-labor regulations under this act. Because you will probably require help only on a part-time basis, and may want to work on weekends or evenings, you might think first of hiring a teenager. Your salvaging project is considered demolition, however, and people under the age of 18 are prohibited under the Fair Labor Standards Act from working on a demolition site.

According to the child-labor provisions of the Fair Labor Standards Act, anyone under 18 may not be hired for hazardous employment. Seventeen hazardous, non-agricultural occupations are listed, among which are wrecking, demolition, and shipbreaking. According to this Act, you may not hire anyone under 18 for any phase of the salvage projects, including cleanup. If you are in doubt about an applicant's age, request age verification and keep it on file for your protection.

You should remember, also, that this is a federal act. States also have laws governing child labor. Your state's laws may be more restrictive than the federal law. In any case, you must always observe the regulations with the more stringent standards, whether they be in a state or federal law.

Also, remember that agricultural and non-agricultural occupations are regulated with different standards according to the Fair Labor Standards Act. Occasionally, a demolition project may qualify as an agricultural occupation. Check with the Wage and Hour Division in your community before proceeding with any project. As a general rule, however, you should not hire anyone under 18 for demolition work.

Minimum-Wage Guidelines

The employment of adults is also regulated by state and federal law. One regulation you should be aware of is minimum wage. As of January 1, 1981, the Fair Labor Standards Act set a minimum wage of $3.35 per hour. Most states also have minimum-wage guidelines. In some states, such as Alaska, the minimum wage is considerably higher than the federal guidelines. In other states, it is lower. Check on the minimum-wage standards for your state. The local office of your state's Employment Service should be able to give you this information. Look under "(your state's name)— State of" in the phone book.

Your demolition project may also be governed by federal minimum-wage law. There are several things to consider:
1. How many people will you be employing?
2. Do you normally have other employees in a retail business?
3. Will the material you salvage be sold across state lines?
4. Are you a farmer who normally employs other workers?

If you are just hiring one person to help you, and the materials you salvage will not be sold across state lines, you will probably not be governed by federal minimum-wage requirements. If the building is part of a larger retail business, if you plan to sell the salvage in interstate commerce, or if you are a farmer with many employees who wants to demolish a building on your property as part of your farm work, then you probably are governed by the Fair Labor Standards Act, and are required to pay federal minimum wage. Before hiring anyone, consult your local office of the Wage and Hour Division.

Finding and Interviewing Applicants

When you have determined whom you can hire and what you must pay for them, you can begin looking for workers. One way to find people interested in this kind of work is to run an ad in the local newspaper under the Help Wanted section of the classified ads. Another way to find workers is to contact your local job service or private employment agencies. When you contact these employment agencies, tell them you need someone with experience in demolition. If you can't find someone with demolition experience, try at least to find someone with carpentry experience. If the person knows how buildings are put together, he will have some idea of how they come apart.

Remember that in many states you will be paying your helper more than $3.00 per hour. The money is an important consideration, especially if you must spend half the time explaining the procedures to him. Check an applicant's experience in the particular job you have for him before you hire him.

Besides questions about experience, also ask the applicant if he has a fear of heights. Some old buildings are very high, with steep roofs. You may want to salvage intact cupolas or other ornate structures from these steep roofs, and will probably need a worker who can handle heights with confidence.

Make sure that you explain the exact conditions of the job and describe the building to the applicant. Dismantling a building is hard physical work. Often, you and your workers will be exposed to sun, wind, rain, and cold. You do not want to hire someone, only to find that he quits when he realizes the extent of physical labor involved.

When you have all the necessary permits and the workers you need, you are ready to begin working at the site.

Safety Precautions

Salvaging a building is hazardous work. Nails protruding from boards, structures collapsing, and debris scattered around the grounds are all reasons why you should use caution when working in the area.

Insurance

Even before beginning work at the site, you should think about safety. Your first consideration should be to get the insurance coverage that is needed. If you are hiring workers, you will need two kinds of insurance: one to cover the workers, and one to cover people who may wander onto the property to see what you are doing. (The number of onlookers you have may surprise you.)

If you already have homeowner's insurance, first call the agent who handles your homeowner's policy. Be sure to discuss your project thoroughly with the agent. As a general rule. your salvage project will *not* be covered under your homeowner's policy. This is true even if the building is on the property normally covered by the policy, so *don't* assume that you are covered under your homeowner's policy. Talk to your agent to learn the extent of the coverage your current policy entitles you to.

Usually, you will have to purchase an excess-liability policy. This policy will provide you with liability coverage in the event that someone wanders onto the property and is hurt. Purchase a policy that will provide you with enough coverage to protect you in any situation (from a person who gets scratched on a nail to someone who is seriously injured or killed while on the property).

The price of the premium may vary depending upon the location of the building. The number of passersby will probably be greater in an urban area than in a rural area. Since this increased traffic increases the risk of accident, you may find that your premium is higher in an urban area than it is for similar coverage in a rural area. We do *not* recommend cutting down on the coverage to save money on the premium. You will find the project proceeds much easier if you do not have to worry about liability suits. Get the most complete coverage you can. If the agent who provides you with homeowner's insurance cannot write an excess-liability policy for you, ask him to recommend an agent who can.

The second type of insurance needed is coverage for your workers in case they are injured. For this coverage, inquire about Workmen's Compensation Insurance. Workmen's Compensation Insurance policies vary from state to state. In Wisconsin, for example, employers with three or more employees or a payroll of at least $500 in any one quarter are required to pay for a Workmen's Compensation Insurance policy. This policy provides unlimited coverage in case an employee is injured on the job. Medical payments and liability suits are covered by this policy.

If Workmen's Compensation Insurance is not compulsory for your salvage project, you can still elect to be covered by this insurance. Your insurance agent can tell you how to secure the policy. If you need further information about Workmen's Compensation Insurance, you should contact the office that deals with Workmen's Compensation in your state. States vary, but your local job service agency can probably put you in touch with the correct office.

When you are buying insurance to cover your workers and people who may wander onto your property, make sure that you are also covered in case of injury. You will probably be on the site more than anyone else, and you should not neglect your own safety in this area.

Posting the Property

A demolition site can be considered an attractive nuisance. Removing boards and windows from a structure is a sure way to attract onlookers. Even though you have excellent insurance coverage, keep as many curious people away from the site as you can. Your salvage project will proceed more safely without interruptions from the curious, and you'll get more work done, too.

Before you begin salvaging the structure, post the site against trespassers. Put "No Trespassing" signs wherever there is easy access to the property, so people who want to step onto the property cannot avoid seeing them. You can protect yourself further by placing a fence around the demolition site. If you have posted the site and placed a fence around the demolition area, you have taken reasonable precautions to protect passersby. Usually, you will not be responsible for an injury a trespasser incurs while he is on your property without your permission. An exception to this general rule may occur when there is a danger on the property, such as a trap you have set up or natural trap that already exists (a sand pit or quarry are examples of natural traps). In any event, the best guideline is to take all precautions available to you. Don't assume that just because you did not invite the person onto your property you are free of responsibility for his safety.

Some municipal codes spell out safety precautions that must be taken before you can obtain a permit to begin demolition. Even if your area does not require posting the site and fencing the area, it is wise to do so.

Illus. 1. Before starting to work, make sure the proper signs are posted. A wrecking permit is required in most places.

Vaccinations

Before you begin the salvage project, you should also make sure that everyone working on the site has been vaccinated against lockjaw. Puncture wounds are among the most usual accidents occurring on a salvage site. Although a nail driven into the hand or foot is most often thought of as a possible cause of lockjaw, any open wound or sore is a potential breeding ground for *Clostridium tetani*—bacteria that grows in the absence of air. This bacteria is abundant in manure and dirt.

Lockjaw occurs if oxygen to the wound is cut off as the wound heals. Since lockjaw can be fatal, you should take every possible precaution. Today, it is generally thought that a tetanus shot provides protection against daily cuts and scratches for about ten years. If you plan to work around a demolition site, however, you may require a tetanus booster even if you have been vaccinated within ten years of the date you begin working on the site.

Before beginning work on the demolition site, consult your doctor. Tell him what you will be doing, and ask him how long it has been since your last tetanus shot. Even if it has been two to five years, the doctor may recommend a booster.

Even if you have had a recent tetanus booster, you should see a doctor if you sustain a puncture wound or any other kind of wound that cuts into the skin, thus providing an entryway for bacteria. In this, as in all other safety precautions, it is best to take a little extra time and prevent serious problems from occurring.

Clothing

Even though you are adequately insured and have recently had a tetanus shot, you should still be careful when working on the site. One way to reduce the risk of accidents is to dress correctly. When I told my father-in-law I had contracted to tear down a barn, he said, "Get yourself a box of bandages and a bottle of iodine. You'll need them." Proper clothing can, however, reduce your need for the bandages and iodine.

First of all, buy yourself a good pair of work gloves. Select a pair with leather palms and fingers. The gloves should come up around your wrists for extra protection. Wear these gloves all the time while you are working at the site. Wear them when you are carrying boards, as well as when you are actually doing the wrecking. Although they may seem hot and cumbersome at first, you will get used to them. Gloves will protect your hands from cuts, slivers, and chafing.

Illus. 2. Long sleeves and leather-faced gloves are important for safety on the job.

18

Slivers and chapped hands are painful, but puncture wounds are even more serious. Yet an injured foot caused by stepping on a nail is one of the most frequently reported accidents on a salvage site. Proper footwear will help protect you against puncture wounds. Thick-soled shoes or boots will reduce the risk of a nail puncturing your foot as you work on the site. Although nails may still penetrate the sole, you will have more time to react. This footwear may seem heavy and hot compared with canvas shoes or sandals, but the protection it gives you is well worth any inconvenience you experience.

As you work on the salvage site, you will kneel on rough surfaces and brush against sharp corners. Protect your arms and legs with a long-sleeved shirt and long pants. This clothing, although hot in summer, will protect you against the painful irritation of scraped and bruised knees and elbows. It has the added advantage of protecting you against sunburn, too.

Workers on a salvage site should also wear protective headgear, such as a hard hat. Even the most careful worker runs a risk of head injury from loose boards or collapsing debris. If you are working alone you know exactly when boards are picked up and carried, but if several people are working together the chance of being hit by a board or falling debris increases. Wearing hard hats while working on a salvage site is a wise practice.

Eye protection is also advisable, especially for certain projects. Wear safety goggles when chipping concrete or plaster; they will help prevent eye injury, which can result if the sharp particles hit your eyes.

Don't allow warm weather to lure you into discarding your protective clothing. If you wear clothing appropriate for your task, you will reduce your chance of accident and keep your salvage project moving along smoothly.

Work Habits

Careful work habits are essential for safety on a salvage site. The first thing you should do is keep the immediate work area as free from debris as possible. Although this sounds like an impossible task on a salvage site, there are some things you can do to reduce risk. Take time to stack boards away from work areas and haul debris to a dumping site. Make sure walkways are cleared of bolts or nails after a column or support wall has been removed. Maybe you will remember to avoid debris nine times out of ten, but the one time you are bent under the largest load you will forget, and end up flat on your face with the load on top of you.

Working with one or more helpers has two advantages. You will not lose time when you encounter large structures that require more muscle power than one person has. You can just call for help and complete the job quickly. Helpers also give you a safety factor. No matter how carefully you work, the chance of a really serious injury is always there. If you are badly cut or knocked unconscious while working alone you might even lose your life, whereas a helper could have driven you to a hospital for emergency treatment. For these two reasons, working with one or more partners is wise.

You will have to pay a small price for having helpers around, though. There is the responsibility of knowing where they are at all times, and also a need for more careful planning before every move. On some parts of the job, especially carrying salvage or tossing boards or rafters down from the roof, you will have to move more slowly.

However, a careless partner is worse than none, because you can be injured by a recklessly thrown board or tool. I have met some careless workers who fling tools off a roof when they are finished with them without even giving a glance below. Discuss the dangers with your partners before you start and call attention to any changes in your location as you are working. If everyone knows the location of the other members of the party, partners will greatly reduce your personal danger, and the job will be completed much sooner.

Illus. 3. Throwing debris down stairs that are used regularly is a dangerous practice. Keep stairs free from any rubble that may cause falls.

Illus. 4. Notice the splinter protruding from the floor on a constant area of traffic. Take the time to remove any objects that may cause falls.

One of the more frequent causes of accidents is carelessly handled rafters, joists, or other supports. These are frequently toenailed to another part of the frame, so when they are pried out, the nail points stick out from the end at a slight angle. The entire timber takes on the appearance of a giant spear, with nails for the spear point. The only way to carry such timbers is with the end first, but the mass of the timber will assure complete penetration of the nails upon anyone they hit.

Walking around a corner into someone carrying one of these timbers could be a disaster, so take three precautions. Before carrying a large timber, always pound the nail points over where they protrude at the end. To further lessen the danger, give a shout as you approach blind corners to prepare anyone in the area. Finally, remember how far and fast the end of a long timber will move if you make a turn. For you, the distance your body moves may be only a foot or so, but the end of a timber 10 ft (3 m) out may pass through an arc of 10 or 12 ft (3 or 3.7 m) at very high speed. If someone walks into you as you turn a corner, the swinging timber can become a powerful club. So move slowly, keep an eye on the ends, and remember where your partners are working.

Another cause for painful accidents is carelessness when using a pry bar. Sometimes you must apply tremendous force to start a joint opening, or even to pry out a rusted spike. The jointer nail may give suddenly, allowing your hands to snap back with that same amount of force. Even hitting a smooth board with your hands can be painful, but if the board has splinters or nails where your hand strikes, these will be driven right through your hand. To prevent this, always examine the area your hands will strike if the pry bar should come free suddenly. Before applying real pressure, flatten any nails your hands may land on. If you follow these two precautions, you will greatly reduce any chance for an accident.

Still another cause of accidents involves using a chain saw. For certain parts of the process, chain saws can be real timesavers, and sometimes the only practical means. But chain saws are dangerous, and unless you have used one in the past this would not be the place to learn how to use one. If you are unfamiliar with the basic techniques, find someone else to do any sawing that must be done.

In addition to the usual dangers of chain saws, you will be sawing timber that may hold hidden nails or bolts. Pull out all you see before starting the saw. Then keep your face well out of the way as you cut, because the saw might strike metal and kick back at any time. A saw with a chain brake is a great safety factor here.

One of the basic rules for using a chain saw is never to reach up to cut higher than your head. You have poor control of the saw, and if it kicks out you could end up falling on the moving chain. If you must cut higher up, be sure both you and your ladder or scaffold are firmly supported. The smallest shift in a ladder can throw you off-balance, and the consequences can be disastrous.

If you wear proper clothing (including gloves and goggles), and follow the rules mentioned above, you will greatly reduce your risk. But cut slowly and concentrate on what you are doing, because the danger cannot be completely eliminated.

Coping with Injuries

Even when you take safety precautions, accidents can occur; therefore, you should have a first-aid kit at the demolition site. You can buy a first-aid kit at a pharmacy. If you are trained in first aid, you may decide, instead, to make up your own kit. Whether you purchase a ready-made kit or assemble your own, the kit should be adequate for your needs.

A kit that is adequate for wounds encountered at a basement workbench may not be adequate to cope with the injuries you could sustain on a demolition site. Whether purchasing or assembling a first-aid kit, consult your doctor about what equipment is necessary for the kind of work you are doing. Helpful information can also be found in first-aid books.

22

Whether you buy a preassembled kit or make your own, be sure the contents of the kit are properly packaged and kept clean. Few work areas are dirtier than a demolition site, so take every precaution to keep your first-aid supplies clean. Also, make sure that the contents are neatly arranged and can be found quickly. If you have to search through the kit for the proper bandages or scissors, one of the advantages (promptness of treatment) of having the kit in the first place is diminished.

By observing basic safety rules such as adequate insurance coverage, proper clothing for the job, and good work habits you will move your salvage project along smoothly.

Illus. 5. This worker is taking a great risk by working with no protective clothing and by cutting over his head.

Tools for Salvaging

Any building is easier to dismantle than it was to build. To the person doing the dismantling, this means fewer tools will be needed. Among the tools *not* needed are levels, squares, plumb bobs, measuring tapes, and anything else used for proper alignment. The tools that *are* used are heavy steel tools which provide a solid reliable grip.

Perhaps 80 percent of the tearing-down process can be handled with just one tool—a wrecking bar or pry bar. These are available in several sizes and styles. Of the many types you may find, one model has proven itself over and over to be superior for nearly every type of dismantling. Rather than being forged in one piece like all other wrecking bars, this one is built from two separate pieces of steel. The pieces are joined at the head by a weld joint.

Knowing the great force often applied to wrecking bars in any demolition project, I was skeptical about the strength of the joint when I first saw one of these bars. After seeing what they can do, however, I am convinced that they are indestructible. Their usefulness comes from three points. First, the two-piece construction allows two separate nail pullers on the head. Each one lies at a different angle to the shaft, so two different types of leverage can be applied to stubborn nails or nails at odd angles. Second, the bottom end of the shaft is far better in its design than that of other bars. The end is wider and thinner, so it can be more easily inserted between two boards. The extra width spreads the prying force over a greater area, so boards are less likely to be damaged than with conventional bars. The third advantage is that the head curvature is less than in any other bar, and the metal wider and flatter. The head can be used as a heavy hammer, saving time in locating another tool. Obviously, these special wrecking bars are worth their slightly higher price tag. They are found only at good-quality hardware stores.

Another tool—one that many people have never heard about—is a nail puller. Designed like long, beak-nosed pliers, a nail puller can neatly reach inside a board, grab a nail by the head, and pull it out with surprising ease. Where the boards are especially valuable, as with choice, weathered barn boards or select sheeting, a nail puller makes very good sense. It can pay for itself with only one or two saved boards.

A heavy hammer or sledgehammer with an 8- to 12-lb (3.6- to 5.4-kg) head is valuable occasionally. Heavy fixtures or timber often move easily once they are loosened or start to move, but loosening them can be a problem. A solid blow from a sledgehammer usually starts most things moving.

A conventional claw hammer has occasional uses, along with screwdrivers and crescent wrenches for internal fixtures. A pair of wire-cutting pliers comes in handy for old electrical wiring and other odd wires that appear from time to time. You may also need an old wood-saw or, for buildings with large beams, a chain saw with an old blade.

Illus. 6. Two different styles of pry bars. The one on the left is the author's favorite.

If you plan to dismantle and salvage plumbing fixtures and piping, you need an assortment of pipe wrenches and a plumber's blowtorch to free soldered joints. A hacksaw is necessary for piping and cables that cannot be dismantled. For cables and flexible conduit, some people use an old axe, which is faster than a hacksaw but more dangerous, not to mention lacking in aesthetics. A cold chisel will often work, too, and offers the advantage of freeing fixtures embedded in concrete.

An old knife, a pair of conventional pliers, and a wood chisel will round out the small tools needed. You will also need a step ladder and a ladder long enough to reach the roof. A wide, flat pry bar with a notch cut from the end and perhaps some 30- to 50-ft (9.1- to 15.2-m) lengths of ⅜-in (9.5-mm) nylon rope will come in handy. With these tools you should be able to tear down any wood-frame building quickly and efficiently.

The tools for tearing down any building are usually the oldest ones anyone has in his tool box. With an occasional spot of rust on them, they camouflage readily with old boards. After a barn I was tearing down permanently swallowed a nail puller and a wrecking bar, I decided something more was needed. A spray can of fluorescent orange paint did the trick. By spraying a few coats on at least part of every tool, you will save countless minutes of searching through the debris for your tools. Spray them all before you start; the effort will repay itself many times over, for the tools will easily stand out against a pile of drab boards or shingles.

Basic Dismantling Procedure

Now that you have decided on the building to be torn down, and have signed a contract so you know your rights and corresponding responsibilities, just how do you begin? The size of one person seems pitifully small when he stands beside a large building. But the building was put together in a logical sequence, and a logical sequence can be used to reverse the process even more easily. Instead of cutting angles and measuring timber, you will be taking apart the fitted angles other workmen assembled.

Although both houses and barns have walls, roofs, doors, and windows, these structures differ greatly in the two kinds of buildings. Houses have both outer and inner walls, while barns have only outer walls; houses seldom have roofs as high as those of barns, with a resulting difference in the approach; house doors slam while barn doors roll; and house windows are usually large and can be opened, yet barn windows are few, small, and usually fixed in position. The kind of fixtures and timber to be salvaged also varies.

Because of these basic differences in their construction, the procedures for taking apart houses and barns are treated in separate sections of this book. Furthermore, because of the greater complexity of houses, the book considers house dismantling in two parts. One part covers the inner structure, including windows, doors, plumbing, heating, electrical fixtures, and cabinets. The second part covers the outer structure, including roofs, walls, chimneys, and all other structural parts of the building. This division should allow anyone, regardless of his background, to follow the sequences with little difficulty.

Dismantling the Inner Structure

The difference between tearing down a large house and tearing down a barn or cabin is considerable. For that reason, these structures will be dealt with separately. We will assume the buildings are in reasonably sound condition and are safe to approach. Otherwise, the danger is too great to consider, and the specialized equipment and techniques become so varied they would require a special emphasis on high-hazard or unusual situations—conditions this book does not advocate.

With the contract signed so all parties know who owns everything on the premises, the real work can begin.

Starting Work

Before anything else, double-check to be sure the electric power, water, and gas lines have been turned off by the utility companies. Then you can work without inconvenience or danger from shocks or explosions. To prevent possible damage to them, remove all the interior fixtures and furnishings first. If carpets remain, pry them out at the wall and roll them up.

You may have to pry off mouldings, and sometimes a base shoe. Often, these can be saved. Mouldings may have special value for restoration projects. If you find old mouldings that may have value, use an old chisel to cut through the paint where the trim meets the wall. This helps prevent breaking the trim, which is usually very brittle if it is old.

The condition of the carpets dictates whether they can be reused. Depending on weather conditions, they can be carried outside, set aside inside the building, or loaded into a covered truck or van. The edging strips can usually be reused. Pry them out from underneath and keep them together in a box or barrel so they don't become lost in the debris.

Another floor-covering material that can be salvaged occasionally is tile squares. Use a wide chisel under the edges to break them loose. You may lose many or none depending on the kind of adhesive used. As soon as they have been removed, stack them in a safe place, because they are fragile unless they are lying on a flat surface.

Interior Fixtures

The kitchen and bathroom fixtures should come out next. For these, you will need a pipe wrench, a screwdriver, probably a hacksaw, and an old chisel. To remove a toilet, flush the unit to get rid of most of the water in the flush tank, and bail out the rest of the water in the upper tank. Use the pipe wrench to disconnect the fresh-water pipe from the tank. At the base of the bowl are a pair of small bolts covered by lift-off ceramic covers. With a wrench, loosen and remove the nuts from the bolts. The entire bowl is now loose and can be lifted up to free the base from the soil pipe.

Some models have wall-hung tanks that must be unbolted from inside. These usually have a connecting pipe joining them to the bowl. Use a pipe wrench to loosen and remove the pressure fittings that hold these in place. As you remove these fixtures, remember that they are usually very heavy, so be prepared for their sudden fall when you disconnect them.

Save all the fittings you remove. Even the bolts that hold down the bowl can usually be used again. New ones are made of plastic and hold up poorly. The older ones made from solid brass are much more durable. These can be removed from the top of the soil pipe by sliding them outward from the pipe. Replace the nuts on them so they are not lost.

Sinks may be styled in a variety of ways, but they are seldom difficult to remove. Start with the drain pipe underneath the basin. Disconnect the "U"-shaped water trap by turning the connectors on both ends with a pipe wrench. The water trap should lift out.

The hot- and cold-water lines leading to the basin are usually ½-in (12.7-mm) or ¾-in (19.1-mm) copper tubing. Using a pipe wrench, you may be able to loosen the pressure fittings that connect the copper lines to the basin. If they cannot be removed, saw off the tubing with a hacksaw.

The last step is to remove the bolts that hold the basin to the wall, if it is bolted on directly. In most cases, the basin rests on a metal bar called a hanger bracket. Lift the basin up, and it should slide free from the bracket. The hanger bracket will be bolted to the wall, so remove the bolts and save the hanger bracket and bolts for reuse.

You may have to remove the hot- and cold-water faucets to loosen or even get at the fittings behind. There are probably more variations on plumbing fixtures than on anything else in a building. But all these fixtures screw together, so they all come apart the same way. About all you can do is keep unscrewing couplings until everything has been dismantled as much as needed to remove the part.

Bathtubs are very similar to sinks, except that in many cases a separate access, usually a closet, opens onto the drain and fittings. The access will likely be from an adjoining room. You may even have to unscrew a false wall panel to reach the fittings.

When you locate the fittings behind the tub, unscrew them with a pipe wrench. Cut off the tubing with a hacksaw if the fittings are corroded in place. You may have to dismantle the drain connections from the basement on some models. Once all the fittings have been disconnected the tub should slide out. You may have to go around the tub-wall joint with an old chisel to free the tub from multiple layers of caulk and paint, but it should loosen easily. If the tub is not cracked or damaged, carry it out where it will be safe, keeping all of the fittings in the tub so they are not lost. Put an old newspaper or cloth under the fittings so they don't scratch up the tub.

Illus. 7. Vise Grip® pliers can be used with a pipe wrench to turn or hold couplings on a water line.

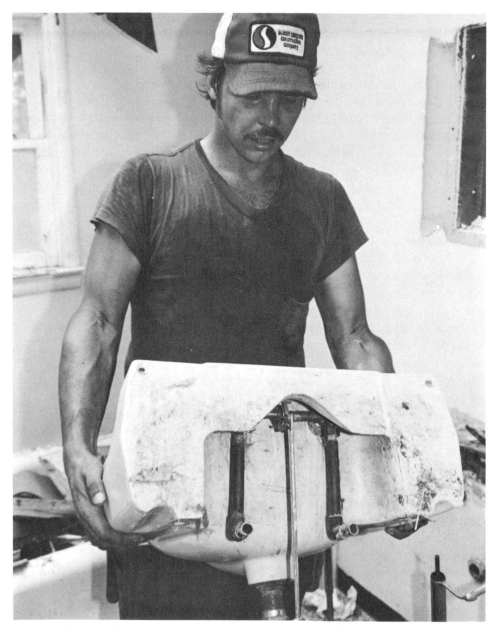

Illus. 8. Once the fittings are disconnected, a sink can be lifted out.

Furnaces

Since heating methods have changed during the past century, the system in any old building may be a complete replacement for the original system, or a series of modifications or additions to the original system. Then, again, you may find the original system intact. The system you find depends on the age of the building and the fuel preferences at the time the system needed repair. Many wood furnaces were converted to oil furnaces as fuel oil became popular. These units are still operating well in many buildings, so they can be salvaged. Old wood furnaces in good condition may also be sold today because heating with wood is popular again.

The common forced-air systems are easy to take apart. From the basement, follow each heating duct from the furnace to its entry point into the floor. Start dismantling the ducts where they change angles to enter the floor. Some of the sections may be joined by a screw that must be removed, but many sections can be forced apart with a blow from your fist. If this fails, try working a screwdriver into the joints to loosen them before you start bending the ducts with a hammer.

Some ducts are round and may be supported by wires or wood slats nailed under the joists. Others may be rectangular, fitted tightly between joists, and nailed on the sides. All are easily removed, because none of the fastenings will be very strong and you will always have gravity on your side. Save all the duct work, because most of it can be used again with only minor length adjustments.

More difficult is the removal of water or steam systems. Rather than light duct work that slides apart, you will be dealing with water pipes, some of them fairly large. You will need pipe wrenches with very long handles to apply enough leverage so the joints can be turned. Since these are expensive and seldom used, rent them for the project.

Each room will have at least one radiator to disconnect and carry out. The larger ones are heavy, so you will need help to carry them downstairs. If you can locate someone who plans to install the same kind of heat, you may be able to sell all the old radiators. They may need flushing out or new valves, but they last exceptionally well.

After disconnecting the duct work or pipes in the basement, carry the furnace itself outside. Some furnaces are heavy enough to require several people, especially with steep basement stairs. When you have successfully removed the furnace, cover it with a tarp to keep the rain off until you can transport it someplace for storage.

Electrical Fixtures

Electrical fixtures are easily removed. You will need a screwdriver, wire-cutting pliers, and probably a pair of conventional pliers. Wall receptacles are removed in two separate pieces. One or two outer screws separate the plastic outlet cover from the box beneath it. Be sure to save all the screws you remove, because fixtures have changed and screws may no longer match. Beneath the plate, screws hold the receptacle in the box. Remove the two screws and pull out the receptacle. Loosen the screws that hold the wire ends in place and pull off the wire loops. The metal outlet box is usually held in by two screws on the inside. Remove the screws and you will free the metal box. Depending on their age, many outlet boxes can be reused.

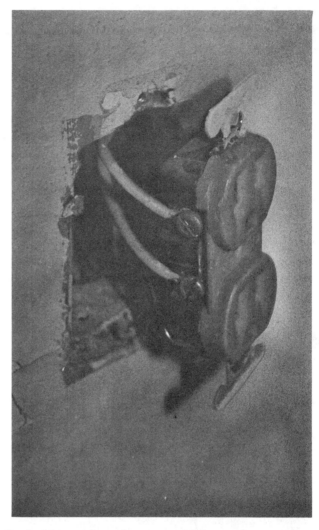

Illus. 9. Once you have removed the outer cover, electric outlets can be unscrewed and pulled out so their wires can be cut for removal.

Light fixtures are usually held in place by a pair of round nuts or a decorative spindle. Removing these will allow the light to drop down to the end of its wire slack. Use the wire cutters to clip the wires, and the unit is ready to use again. The box beneath should come out the same as the wall outlet boxes.

The main fuse box and connections probably cannot be reused unless the building is fairly new, because wiring codes have changed over the years. Older fuse boxes have been replaced by boxes containing breakers that only need resetting when circuits become overloaded. These have made fuse boxes obsolete. To remove the box, look for bolts that hold it into the wall at the top and bottom. By removing these, you can lift off the cover and expose the bolts inside that hold the lower part of the box in the wall cutout.

Either above or below the box a metal pipe enters, held into place by a retaining nut. Loosen the nut with a wrench, and cut off the electrical cables inside the box. The box can then be pulled out of its wall cutout.

Sections of metal conduit will only be in the way, so now is a good time to remove them. Cut them with a hacksaw wherever they enter a wall. Some of them can be reused, along with the conduit straps holding them to the wall.

Outside, where the wiring is grounded, you will find a buried ground rod. A few shovel strokes will free this for reuse.

If the house has a doorbell, remove it for reuse. The unit has a cover that pries off to reveal the screws holding the unit on the wall. Clip the wires leading to it and remove the screws, saving all the parts to aid in reinstallation.

Many houses have added electric baseboard heaters since oil prices began to rise. These can always be used again. Remove the outer cover to reveal the wires below. Use a wire cutter to clip off the wires inside the base. Remove any screws that hold the unit against the wall, and lift the unit out. Be sure to save any thermostats that may have been added for individual room-control. These are always in demand. Again, save all the mounting bolts or screws and keep them with the unit so they are not lost.

Wall thermostats, too, can be resold or used again directly. Pry off the cover with a screwdriver. Inside you will see two or more small screws that hold the unit to the wall. Remove the screws and pull the remaining part out from the wall. Use a wire cutter to clip off the small wires. Then clip the two pieces of the thermostat together and set it in a safe place; jarring it might cause a loss of accuracy or damage the thermometer.

Cabinets

Interior cabinets, cupboards and counter tops should come out next. Nearly all of these can be used again. If they are in good condition, they may be useful as kitchen cupboards exactly as they are. If they are in poor condition, they can still be used in the basement or garage to hold extra storage.

Smaller cabinets can be removed in one piece, but larger or built-in units may need to be disassembled at least partially, both to help remove them from the wall and to allow them to be removed from the building. Begin by taking out all the drawers and doors. Drive out the hinge pins on doors by using a hammer and an old screwdriver. Many of the pins will be locked into place by many layers of paint, so you may have to chip away the paint to uncover the joint between pin and hinge. Keep all hinge pins in one of the drawers so you can reassemble the unit at a later date.

If the doors have recently been painted, or the paint is in good condition, put paper or cloth between the doors wherever you stack them, to keep the paint looking good. The more beat-up anything is, the more difficult it is to sell as salvage.

Past this point all cabinets are different. You will have to decide as you proceed how much to disassemble. In every case, remove from top, bottom, and sides all trim-moulding strips. These may be partly blocked in at the top by a suspended ceiling added after the cabinets were installed. If so, break away some ceiling tiles to reveal the moulding. If you think the moulding can be used again pry it off gently to prevent breakage. If it is painted heavily in place, use an old chisel and hammer to crack the paint where it meets the wall or ceiling. Then pry it off with the long end of a wrecking bar or, if it is thin, the chisel.

Illus. 10. A pry bar finishes the job started by an old chisel, which was tapped into the joint between the outer facing piece and the cabinet to break the old paint. Pry the facing wood slowly, so the old wood has a chance to give slightly. This prevents breakage.

With all mouldings off, look inside the upper unit to see how it is held. It may be lag-bolted to joists in the ceiling or studs in the wall, or may instead be nailed to the studding. If it is lag-bolted, use a crescent wrench or socket wrench to remove the bolts. With nails, use the wrecking bar to pry outward on the back. You will probably find only plaster walls to pry against. These will only cave in, offering no support for the bar. Slide a flat piece of wood between the bar and the wall to spread out the pressure on the wall. You can exert more force this way.

Illus. 11. This old cupboard with trim mouldings removed can be pried away from the wall.

If the unit will not come out, you will have to start disassembling it. As you pry off each board, use a soft-tip marker to label the pieces for position. Otherwise, you will waste too much time trying to figure out where the parts go when you put it back together. If the cabinet has a large number of small doors and drawers, label these as well. Especially in very old buildings, almost every part was custom-fitted for one place, so no parts will be interchangeable without planing, sawing, or filing.

The lower unit may have a counter top that is in better or worse condition than the rest of the unit. The counter top can be pried off with a wrecking bar if you plan to discard one of the two parts. If the unit is large and heavy, though, the top often makes a convenient carrying handle, so leave everything intact until you have moved the cabinet to its final position. Pry outward on the back of the cabinet to loosen it from the wall, and upward on the base if it is nailed down. When it is freed, carry it and all shelves, doors, and drawers out of the interior where they will be out of the way. Store them under a tarp or in a storage shed so they will not be exposed to the weather. Rain-soaked drawers seldom fit without endless planing.

Tiles

Ceramic tiles are as popular today as they have ever been, so if the walls, floors, or counter areas have been tiled the tiles are well worth saving, unless they are cracked or badly discolored. They are easy to remove once you can get at an edge where they join some other surface material. Look for a trim moulding that joins the edge of the tile to a wall or floor. Pry off the moulding to expose an edge of the tile. Then use a wide cold chisel and hammer to force the tiles up from underneath. They are much faster to remove than to install.

Store the tiles in a cardboard box until you have a few extra minutes to clean off the old glue or grout that adheres to the edges and bottom. A cold chisel and hammer work well in this process. The cleaned tiles will be all ready for use on another project.

Towel Racks

Towel racks have always taken many forms because people like to show their originality with them. You may find commercial models selected to go along with the room decor, or you may find handmade, one-of-a-kind towel racks built from old deer antlers, rolling pins, or farm implements.

While nearly all towel racks can be used again, those with unusual construction or creative handmade parts may have special value. They are always screwed or bolted to the wall or a cabinet, so all you have to do is find the screws and remove them. To keep all the parts together, use masking tape to wrap the parts securely, including any odd bolts or screws. Then you can either sell the entire unit intact or reinstall it easily.

Doorknobs

Over the last few years, old doorknobs have surprisingly developed a value to collectors. Stylish older buildings are sure to have their share of ornate brass or glass doorknobs. These were built to last, so usually the knob and entire lock mechanism will still work perfectly after a century or more of use. Replacements for lost parts may be impossible to find, so here again wrap masking tape around all parts to keep them together.

The strip located on the doorjamb will lift out after you remove the screws holding it against the jamb. With the door open, you can easily see the screwheads, so use a screwdriver that fits them exactly. If they are corroded into place, you will have to apply more power; a screwdriver bit in an auger or brace is best. With the extra leverage, you can twist out the most stubborn screws. If you lack a brace with a screwdriver bit, try a Vise Grip® pliers on the shaft of an ordinary screwdriver for increased leverage.

The latch mechanism is held in place by screws placed into the edge of the door next to the latch. After removing the screws, pull the latch mechanism straight out.

Several different designs have been used throughout the years to hold doorknobs and lock plates together. In the oldest models, a slotted set-screw holds the knob onto the center shaft. After removing the setscrew, you can pull or pry off the inside knob. The other knob pulls through from the opposite side.

With newer models, the base of the inner knob may be covered by a decorative flange that must be pried off to reveal two slotted or Phillips® boltheads that hold the unit together. You may have to rotate the flange to locate the boltheads, or align a hole to get at them. Remove the bolts, and the knob assembly will pull out, half from each side.

Older locks have a large metal plate covering each side of the mechanism. These usually unscrew from the inside to reveal the lock mechanism beneath. If the knobs and plates are especially ornate, it may be worthwhile to have a locksmith make a new key if the originals have been lost. Keys are often missing for older locks because new locks are added to some part of the door while the old mechanism remains in place, but inactive.

If the door itself is an original and still solid, you may choose to leave the lock mechanism and doorknobs intact, and sell or reuse the entire unit. The choice depends on the condition and quality of both lock and door.

Doors

At some point during the dismantling of the interior fixtures, you will want to remove all the doors. If you remove them early in the process, you will find it more convenient to carry heavy fixtures and lumber through the rough openings without having to get out of the way of swinging doors. On the other hand, theft at demolition sites is a common problem, so you may want to leave all exterior doors in position so the building can be locked up when you are not working. You may find it best to remove all the interior doors before removing flooring and large fixtures from the premises. Then you can remove exterior doors when you take out the windows. After that, anyone can easily enter anyway, so locked doors are of no further use.

Doors are usually easy to remove. Use an old screwdriver or cold chisel and hammer to tap the hinge pins out of the hinge joints. Start with the lowest hinge and work upwards so gravity will not cause the top to kick out, pinching the lower hinge pins. The pins are always inserted from the top of each hinge, so drive them upwards with the screwdriver and hammer. Sometimes they are frozen in position by multiple coats of old paint. To free the pins, tap straight in with the screwdriver at the base of the flange where the pins widen. Once the pins are loose, drive them upwards. In extreme cases, you may have to fasten a pair of Vise-Grip® pliers to the top, and twist the pliers or tap them with a hammer.

The hinges are easily removed by taking out the screws that hold them to the door. You may, though, want to sell or use the door with its hinges; in that case, leave the hinges intact. Reinsert the hinge pin and tape it to the hinge if the fit is loose. Matching hinge pins are nearly impossible to find for old hinges. If the hinges are intact, many people buy them as decorator items.

Door Frames

Door frames are not hard to remove if you reverse the process used to install them. The entire unit consists of nine pieces of wood. Flat pieces that form a border around both the inside and outside of the opening are called casing or trim. There are six separate pieces, three on each side. Begin by removing these. Starting about an inch (2.5 cm) from the floor, drive a thin pry bar or an old chisel into the joint where the casing meets the wall. The casing is only held together with small finishing nails, so the inner edge will usually lift out a half inch or so. Work upwards to the top of the frame, prying gently so nothing breaks.

You will probably find more finishing nails on the other side of the casing, so slide the pry bar farther under and continue prying. To prevent damage to the casing, it's a good idea to place a small wood block in the opening and pry against it. Continue the process to remove the other two casing pieces on that side and the three on the other side of the opening.

You will now see a space between the frame and the wall. Also visible are the finishing nails that hold the frame in place. There are several ways to remove the nails. One way is to use a flat pry bar with a "V"-end to shear off the nails by hitting the bar with a hammer. Or you can remove the wood stops on the inside of the frame, revealing the nailheads that have been set below the surface with a nail set. Use a knife or chisel to cut out some wood around each nailhead so the heads can be grasped with Vise Grip® pliers and removed. Or, if you have a nail puller you can quickly jerk out the nails.

After the nails are sheared or removed, the entire frame unit will slide out. The casing and frame is now ready for reuse. At the current price of these units, door framing is well worth saving.

Illus. 12. Removing all the mouldings is the first step in taking out a door frame.

Illus. 13. With the mouldings off and the facing nails removed, the frame will come out if you pull hard from the outside. This door frame or jamb is ready to use again.

Decorative Mouldings

There is a demand for old, decorative trim mouldings. Some of these were cut from oak in widths of 5 in (12.7 cm) or more to join ceilings and walls. While the originals were usually finished with only stain and varnish, most of them have since been painted over, often with numerous layers of paint buildup.

To remove these heavily painted mouldings, you will need a wide chisel and hammer. Position the chisel at the junction between moulding and wall or ceiling, and tap lightly with the hammer to break the paint joint. To prevent breakage, follow around the borders of the moulding until all the paint is broken. Then pry outward on both sides alternately with an old chisel or a thin-edged pry bar. As you pry, allow time for the nails to release their hold; this way, you can avoid damage to the moulding.

When you have removed all the mouldings from one room, pile them together and wrap a layer of fibreglass tape around each end of the pile, making one convenient pile to be sold or reused later.

Hardwood Flooring

Despite the market's numerous offerings of plush carpeting in every color of the rainbow, hardwood flooring still retains its popularity. There is something natural and comfortable about a varnished hardwood floor set off by a colorful area rug. If you are fortunate enough to be tearing down a building with hardwood flooring, you should remove it carefully, to damage as little of it as you can.

Hardwood flooring is installed in a tongue-and-groove pattern that locks each board to the next, and keeps it level. Each board has both a tongue side with a raised hump, and a grooved side with a channel cut into the board. The first board is usually installed with the grooved side flat against one wall. A short section of flooring is set next to the tongue side and stuck with a hammer to butt each board tightly against the one before it. Then nails are driven at a slant through the back of the tongue down into the floor.

Because of this installation, it is necessary to remove hardwood flooring starting with the last board laid. Otherwise, you will have to work much harder, and the loss in split boards will be very high. The first step is to determine from which side the boards were laid. You may find enough space between the first board and the 2-in (5.1-cm) plate along the sides to see if the board has a groove or a tongue. Usually, too, the last boards to be installed were not the exact width needed, so they were sawed or planed to fit. If you find narrower boards along one wall, you will know that the other side was laid first.

The first board is often very difficult to remove because of its tight fit against the plate. The easiest way, of course, is to wait until the roof and walls have been removed, leaving the flooring much easier to get at. If you wait, however, the flooring might be damaged by rain, swelling, and warping and will be very hard or even impossible to use again. So the best advice is to remove the flooring immediately. Use a fine-ended pry bar, driving it between the plate and the first board with a hammer, and then prying outward using the plate as a lever. Break in the wall behind to allow room to pry.

After the first board is out, the rest are easy. Don't try to hurry, though. Slide the pry bar under the tongue edge of each board, prying upwards slightly and working the bar back far enough so you can pry it up and

Illus. 14. Prying out hardwood flooring is a slow process. A pry bar is the only tool for the job.

over the tongue of the board behind without breaking anything. A few sections along the bottom groove may splinter off, but this will not hurt the boards for reuse. When you have taken them all out, store them where they will stay dry.

Pumps

The tools you may need to remove a pump are two large Stilson® or pipe wrenches, a screwdriver, a hammer, and a hacksaw. Start by locating a union between the pump and basement wall; this union is usually a large hexagonal nut joining two sections of steel pipe. Put the large pipe wrench on the union so you can turn it counterclockwise. You may have to tap the end of the wrench with a hammer to free the union so it turns. When it is free, turn it until the union is completely separated from the fitting that screws into it.

You will now have a length of pipe towards the pump that fits into some kind of coupling on the other end. Twist the pipe slightly away from the union you have just unscrewed so you will have room to unscrew the other end. To unscrew this section of pipe, use two pipe wrenches. Place one on the next coupling so it can't turn and the other on the pipe, about 2 in (5.1 cm) from the coupling.

Remember, with pipe wrenches the pressure should always be applied towards the lower jaw. The wrench on the coupling should face you, while the one on the pipe faces away. With a bit of effort the pipe should turn, unless it is heavily rusted in place. Try tapping the lower wrench with a hammer to start it moving. Then unscrew it completely. You may have to tighten the union on the other end farther onto the pipe to provide space enough to remove the section of pipe.

If the pump is a deep-well model, you will have two pipes to disassemble. Remove all the other couplings the same way. You may find two sections of pipe joined by rubber sleeves. These are held in place by ring brackets tightened by slotted bolts. Use a screwdriver to loosen the screws and the brackets will slide off, or simply cut the rubber connector with a knife, leaving the brackets on until later.

With the two steel pipes removed you should find a copper pipe joining the pump and the water piping in the house. The easiest way to disconnect copper piping is to saw the pipes with a hacksaw. Then use a pipe wrench to unscrew the fitting into the pump.

The pump should now be free from all connecting lines. At the base of the pump you will usually find a pressure tank. Twist the pump counter-clockwise to unscrew it from the pressure tank. Then open the valve on the bottom of the tank to release any water inside. When it is empty, carry the tank, pump, and all piping to a place where they will be safe. These are valuable items because they can be used again.

Illus. 15. Some pump fittings are joined by rubber hoses secured with clamps. Loosen these with a screwdriver.

Water Heaters

Water heaters have a much shorter, useful life span than any building, so if a water heater is left in a building to be torn down chances are very good that it is the second or third replacement—and it may have useful years left. Open the valve at the bottom to drain out any remaining water. The water flowing out will probably be rusty, but this is not an indication of the condition of the unit. While the water is emptying, use a hacksaw to cut the two lines of copper tubing that connect the heater to the water source and the house. If the pipes are steel, their small diameter will make them easy to unscrew at the fittings with a pipe wrench. Cut the electrical connections with wire-cutting pliers. If the unit is gas-operated, you will have one extra set of pipes to disconnect with the pipe wrench. Then the heater can be removed and stored for later use.

Stationary Tubs

Most older buildings have stationary tubs, either for a basement clean-up area or as part of a laundry room. These are very popular today, and easily sold if you don't need them yourself. To remove them, all you need is a short-handled wrench big enough to turn the fitting at the back where the hot- and cold-water pipes connect with the faucets. Turn the connecting nuts counterclockwise until the inflow lines are free.

The drainpipe requires a standard pipe wrench to unscrew the large connectors holding it to the wall pipe. If the tubs have an open front, reach up from underneath. Some models are encased in a metal cabinet extending all the way to the floor. These have a metal panel in front or on the side that swings open to provide access to the drain pipes. With all pipes disconnected, the entire unit will slide out from the wall.

Interior Masonry

In most older homes, fireplaces are the source of heat. Some fireplaces are brick, while others are various types of stone. Some of the stone fireplaces have especially attractive stonework that can be saved to build another fireplace. To prevent damage to the stones, remove them before starting to tear down the structure of the building.

The stonework or bricks you see when looking at the fireplace are not structural supports for the chimney. Instead, they are only a decorative facing. If you want to preserve the facing bricks or stones intact, start by removing them at the mantel, which nearly every fireplace in an older home will have. If the mantel is wood, use a small wood block to tap it upwards from below. It should not be fastened down strongly, and should come free to provide a fine mantel for use over another fireplace.

If the mantel is inset in brick or stone, use the cold chisel and hammer to cut the mortar from blocks above it. Start with the highest row and work down until you reach the one above the mantel. When you start to remove the last layer above it, ask a helper to hold the mantel so it can't fall. Some mantels have been set in only slightly, so both the layers above and below them are needed for support. After the layer above the mantel has been removed, the mantel should lift out.

If the mantel is a stone slab, you will have to break the mortar loose below the slab. You can break mortar joints with a hammer and cold chisel. But if you plan to disassemble an entire fireplace, it might be easier to rent an electric chip hammer from any rental outlet. These cut mortar quickly with far less effort than a cold chisel and hammer. Stone mantels are often extremely heavy, so don't try to take down a large one without help.

With the mantel off, continue cutting out mortar and removing the stones or bricks. Ignore all mortar that clings to the stones or bricks because cleaning mortar is easier to do all at once. If you decide to sell the bricks or stones, you can offer them uncleaned for a slightly lower price. People are often delighted to find a bargain, and prefer to clean their own bricks.

After you have removed all the bricks or stones down to the hearth, carry it outside where falling debris will not damage it. The rest of the chimney, along with the firebox, should be left until later.

If the hearth extends into the room, or decorative stonework or brick-work is raised above the floor for seating, remove all of it so it is not damaged. Some fireplaces have floor-to-ceiling decorative brick- or stone-facing. Start at the ceiling, chiselling the mortar away at the ceiling joint. With the first row of brick or stone out, the next rows will be easier to remove. You may find wooden mantels inset into the stonework. Usually, only the layer of stones above and below hold it in place. When the row immediately above it is removed, it may be loose enough to fall out, so be prepared to catch it if that should happen. Most wood mantels are dried, so even the larger ones seldom weigh as much as stone mantels. With especially large ones, though, it is wise to have a helper on hand.

Illus. 16. Old brick salvaged from fireplaces or brick veneer walls is very popular in new construction.

Dismantling the Outer Structure

Windows

Windows are one of the highlights of any architectural design. Designers usually use them to their best advantage, resulting in a great variety of styles and constructions. Most are relatively easy to remove if you follow a logical sequence.

The general style popular with builders 60 to 100 years ago—the period you are most likely to encounter—consisted of two halves, a fixed upper half and a lower section that could slide up and down. Because these were often very heavy, a series of weights suspended over a pulley inside the wall provided help when the lower section was raised.

To take out a window of this type, start by removing the interior moulding-strips that hold the lower window in the frame. It may be heavily painted into place, so use a small chisel to break the joint open wide enough to insert a pry bar. Start at the bottom and continue prying all around on both sides. With the moulding off, the lower window should lift inward away from the frame. However, it may also be attached by the ropes and weights in the wall. Pull the weights up and disconnect the ropes from the top of the window. Then set the window aside.

The ropes and weights both have uses if you want to save them. The weights have often been used for boat anchors. They can also be used again for their original purpose—to help open large windows on sun porches or open large doors more easily. The ropes are fairly stiff and make excellent jump ropes for youngsters.

Next, pry off the interior mouldings that hold the upper window in place and lift the window out. Set it with the other one where it will be safe. Pry off the windowsill and any other decorative mouldings inside. Then remove the exterior trim on all sides. You may be able to do this by reaching from the inside, but tall windows may require a ladder from outside. With the exterior trim removed, nothing should be holding the frame in place but friction. A few taps with a hammer should free the frame in one piece so it can be pushed towards the outside. Store it with the windows so the entire unit will be ready to use again. Even windows of unusual shapes, such as those found in old churches, are used in hunting and fishing cabins, garages, or outbuildings. Sometimes the more unusual shapes are considered conversation pieces and are sold before the more conventional styles.

Illus. 17 (upper left). On the inside, pry off the trim mouldings that hold the window in place. Illus. 18 (upper right). The window can be lifted out on the inside. Illus. 19 (below). The window may be held by a cord fastened to weights in the wall. Slide the cord out of slots in the side.

Illus. 20 (above left). To begin salvaging a window, pry off the outer trim with a pry bar. Illus. 21 (above right). The upper window lifts outward. Illus. 22 (below left). The outer framework can be lifted out. Illus. 23 (below right). Once lifted out, carefully lower it to the ground.

An older building slated to be torn down may often show touches of elegance. One sign is etched-glass windows. You can recognize them by their frosted effect, usually with geometrical snowflake designs or, in some cases, even figures of animals or complete scenes. Etched glass is in demand because of its beauty and antique value.

The best approach is to keep any special glass panes in the door or window where they were originally mounted. This is usually the best safeguard against breakage. In some cases, though, you may want to sell doors or windows separately from their etched-glass panels. Use care when removing them. Start on the inside and pry off the trim moulding that holds the glass in place. Do not place the chisel or pry bar between the glass and wood. Instead, place the tip on the outer edge where the trim moulding meets the wood of the frame. Tap it in gently with a hammer, lifting the moulding out from the frame. This procedure will minimize any pressure against the glass that could result in a crack. Remember that glass becomes more brittle with age, so avoid all pressure against the panes.

When you are removing the pieces of trim, keep one hand on the pane so it doesn't fall out. Years ago windows were not sealed with caulk, but depended only on careful fitting to seal out air currents. If the pane is large, ask someone to stand on the inside to catch the pane. Then, from the outside, push gently along the top edge with your fingers. The pane should slide out from the top, allowing your helper to catch the pane and lift it out.

Needless to say, these panes are extremely fragile. The ideal approach is to find a buyer for the pane before you remove it, so he can be on hand to pick it up as soon as you remove it. If you must store large panes for later use, protect them carefully and take them home. One means of protection is to encase the panes between two sheets of 1-in (2.5-cm) plastic foam, one on top and one underneath, and wire the ends together so the glass cannot break while being moved.

Occasionally, old buildings may contain stained-glass windows. The technique for removing these is basically the same as that used for etched glass, except that the panes are far heavier, often more valuable, and equally fragile. Plan on a helper for all but very tiny windows, and take every precaution with the packing technique described for etched glass when you are transporting them.

When you have removed the windows, you should find nothing inside the building that is likely to be damaged by falling debris. You are now ready to tackle the structural part of the building. You should have removed all possible obstacles, such as tree limbs or shrubs, so the access is clear.

Now is the time to decide whether to start with the interior walls or the roof. Each holds advantages and disadvantages. If you do the interior walls first, you will be sheltered by the roof. The expected weather and time of year may be important considerations here. If you remove the roof and ceiling first, the interior will have much better ventilation. In some buildings, the dust from old plaster and firewood particles can cause problems for the toughest respiratory system. Without the roof, though, you may be unable to work on the inside walls during rainy weather. Perhaps the best approach, especially if you are on a tight schedule for the project, is to start on the roof, and if rain develops, move to the interior walls until the weather improves.

Eaves Troughs

The condition of the eaves troughs varies from nearly new to bottomless pieces of rust. New ones or those with a few rusted areas can be used again. Rusted areas can be repaired with screen mesh and auto-body putty. Smaller areas can be repaired with metal-repair tapes or Liquid Steel® in squeeze tubes. The troughs fit into metal brackets that attach to the eaves. Pry off the brackets and save them.

Downspouts are often lengthy, their sections joined by rivets. These are sometimes difficult to remove. Use a hammer and an old chisel to raise the edges of the rivets until they can be cut off with side-cutting pliers. Then pry the overlapping metal sections apart. Save all pieces, since most of them can be used again.

Coping with Insects

You will often run into hornets or wasps that have selected the eaves of an old building as a place to nest. If you see their papery nests under the eaves, you will know immediately that you have a problem.

The best way to eliminate nests is to wait until all the insects are sure to be in the nest—either early in the morning or late in the afternoon. Approach the nest slowly and quietly from a ladder. When you are within 2 ft (.6 m) of the nest, open fire with a flying-insect repellent spray. Any insects that fly out through the spray will lose all interest in attacking you. They quickly fall to the ground, where they die within a few minutes. Those that stay inside the nest are destroyed by the vapor soaking through the nest. In a few minutes, you can knock the nest down with a pole.

Another method is to use a smudge on a long pole. Wrap a few layers of rags around the end of the pole, dip it in fuel oil or kerosene, and light it. This method is effective, and completely destroys the nest and the insects inside. There are two cautions, though. The old boards under the eaves are highly flammable, so be ready with a fire extinguisher in case the boards should catch fire. Also, remember that insects have a keen sense of smell when it comes to detecting smoke. They may come pouring from the nest if you are not fast enough at putting the smudge directly under the entry hole, and you will be left with some angry insects.

The most dangerous group of insects I have ever encountered built no external nest, but chose instead to nest inside the eaves between the soffits and roof boards. When I say they were dangerous, bear in mind that I have kept honey bees for a number of years. I thought I was used to sting-ing insects, but these wasps were quite different. One of the guards dropped out and attacked me over 20 ft (6.1 m) from the hive with others close behind him. I was stung three times before I fled out of range, so I developed a healthy respect for them. I put on a head net and gloves before dealing with them further.

I waited until evening and climbed up a ladder towards their entrance hole. As the guards attacked me, I shot out a spray of repellent that drop-ped them to the ground. Then I sprayed the opening, pausing every few seconds to allow flying wasps a chance to climb out and fall to the ground. The spray inside finally killed the last of them, and by morning the ground was heaped with dead wasps. The eaves could now be approached safely.

If you encounter a similar problem, rent or borrow a bee suit with head net, so you can approach the hive safely. The damage a large swarm of wasps or hornets can do is not worth the risk of dealing with them unprotected.

Safety Procedures for Roofs

Before starting on the roof, you will need some means of staying on it, especially if the pitch is steep. You will encounter no danger from nail points when you are removing the shingles, so this is a good time to put on that old pair of tennis shoes or any shoes with soft rubber soles. These offer a better grip on sloping roofs.

For extra security, you should provide a support for your feet. The tiny stone particles that fall from shingles can escalate into a ball-bearing race-way, causing you to slide down the roof. To prevent this, arrange a horizontal board as a footrest. The best arrangement, used by roofers, consists of a series of metal brackets that hold a plank out from the roof at a comfortable angle. You may be able to borrow or rent these, or even buy a set and resell them when you are finished. Lacking these, you can get by with a section of 2 × 4 (5.1 cm × 10.2 cm) spiked to the roof, al-though on a large roof the strain on your ankles may become uncomfortable.

Shingles

It is possible to salvage asphalt shingles if you are willing to spend nearly double the amount of time removing them. Of course, this applies only to shingles in good to excellent condition. Older shingles are not worth saving. If you plan to save the shingles, wait for a warm, sunny day. By midmorning the shingles will flex enough so the interlocking edges can be slid apart without tearing or cracking. But you will have to work slowly, to give each shingle time to flex.

Even if you don't plan to save shingles, observe the precaution of working on a roof only in dry weather. If winds become gusty, stop work. Even the best roofers are sometimes caught off balance, and a sudden gust could cause a fall with disastrous results. When moving around or changing positions, always move more slowly than normal. Keep reminding yourself of where you are so you don't make any foolish moves. Keep all tools stowed under control when you are not using them, because they can slide down a roof and quickly strike someone below. Shingles carelessly tossed off can fly a surprising distance, and do damage to someone's head. People have a natural curiosity about demolition projects, so expect people to stop and watch as you work. As long as you plan on people below, you should never have a problem.

Before you start climbing your ladder, be sure the feet are properly braced. Workers are often careless about that, setting ladder feet on old bricks, boards, shrubs, or debris. A central location is best, unless the footing is more uniform on some other part of the roof. If the roof is high, anchor the top of the ladder by tying it through a window to a 2×4 (5.1 cm $\times$ 10.2 cm) stretched across from the inside. Then the top will never kick out when you come down. If you are especially nervous about heights, fashion a safety harness from a ½-in (12.7-mm) nylon rope. Loop it around your waist, shoulders, and chest so a stop won't hurt you. Then tie a straight section of rope between the harness and a support on the other side of the house or the chimney. You will lose some manoeuvrability, but for anyone afraid of falls this might be the only way to do the job without hiring someone else.

With all safety factors well in hand, start on the roof ridge. If the ridge has a metal cap, pry it off with a pry bar. If instead it has a shingled ridge-cap, start on the end where the last capping shingle was installed and pry off all the capping shingles.

What you do from this point depends on the condition of the roof, the number of layers on it, the materials used for roofing, and your own decision about saving the shingles. Here are some methods to consider:

If you are saving asphalt shingles, you can start at either end of the ridge. The best tool for this is a thin, wide bar with a "V"-notch cut from the end for pulling nails. Force the end under the shingle directly above each nail and pry up slowly. Most roofs will have either two or three nails at the upper edge of each shingle. Once these have been removed, the shingle will be held only by its interlock with other shingles around it. Use great care in separating the overlapped or interlocked edges so the shingles remain intact. If the shingles are very warm and you work slowly, you should be able to save nearly all the shingles. Do one row all the way across, then start on the next row.

The shingles you remove from a large roof will weigh much more than you might expect. If you carry the shingles down the ladder in stacks as you remove them, you will make many trips up and down the ladder and consume more time than most people care to spend. The easier approach is to spin each shingle off the roof like a Frisbee®. Nearly all of them will land flat and undamaged if the area below is reasonably smooth. Try both techniques to determine which one works better for you.

Illus. 24. If you plan to save asphalt shingles, use a wide pry bar to loosen them from the top of each shingle. On a warm day, you should be able to save most of them.

Continue removing shingles in rows until you reach the eaves; then start on the other side of each ridge and work downwards. Whenever you have exposed a section of metal flashing in the valleys, pry the nails out with the pry bar. Often, metal flashings can be used again. Slide them off the roof flat, so they are not bent when they land.

If asphalt shingles are very old and badly deteriorated, their removal is far easier. Start at the ridge, but instead of a pry bar use a square-ended shovel or a pitchfork with wide, flat tines. Scoop under the old asphalt underlayment and roll the shingles in front of you as you go. If the shingles are to be hauled away from the site, park a truck below the eaves and roll the shingles into the box. This way you will not waste time gathering shingles.

Illus. 25. When roofing material is not worth saving, use a square-ended shovel to cut off all layers of asphalt at once.

You will occasionally encounter a roof originally covered by cedar shakes, but later reshingled with asphalt shingles right over the shakes; this poses more of a problem. The wide, thin bar tends to get stuck on the shakes when you slide it under the asphalt shingles, and the strength of the combined layers usually prevents you from prying off both layers at once. Remove the asphalt shingles first, usually with the wide, thin bar, and work in rows. When the asphalt is all off, deal with the shakes. These are applied with a vast quantity of tiny nails that prevent the shakes from splitting. Here again, a wide, thin bar works best. Force it in quickly at the bottom edge of the shingles and they will quickly loosen, often flying off in a shower of dust. After all those years of drying, they are brittle.

Keep an eye out for exceptionally wide shakes in good condition. These are popular with artists and hobbyists for painting scenes or decoupaging colored pictures or sayings. They make fine wall plaques or, with shaped cutouts, unusual picture frames. If you salvage wide ones, set them aside where they will not be stepped on or damaged by other debris.

Illus. 26. Cedar shakes are most easily removed from the bottom with a wide pry bar.

Roof Boards

The roof boards under the shingles were usually pine-planed to a ¾-in (19.1-mm) thickness. In tearing off roofs, I have sometimes encountered boards much wider and of better quality than most available today.

More often, though, the roof boards were chosen for that purpose because they were too poor in quality for anything else. They may have large knotholes or other defects that limit their use, but unless they have been damaged by water they can still be used again. Some of these are long, so a few cuts to remove damaged areas will still leave them long enough to be useful.

Start at the roof ridge. Use a long wrecking bar with a well-sharpened end. Jam the sharp end between the roof board and the rafter wherever you find a nail holding the board down. Pry up at each of these, raising the board a short distance all along its length. When you have raised the upper edge, you may find the easiest method is to reverse the bar and use the hook end to continue prying. If nailheads pop up, use the hook end to pull out the nails. Otherwise, leave all other nails until later.

As the boards begin to accumulate, you will have to remove them from the roof. If the roof is not too high, slide the boards flat over the eaves and let them fall on end. This technique will be least likely to cause any damage to the boards. As you remove the boards, you will also be removing your footing. If you dislike balancing on the rafters, save a couple of boards to stand on, sliding them down the roof with you as you go. A spike driven in halfway will hold the boards temporarily in place. You can also stand on the boards below, removing boards above you as you move down the roof. You will have more secure footing, but the prying angles are not as good as the angles you have while on the boards. Continue removing boards down to the eaves; then do the remaining portions of the roof the same way.

The approach described above applies only to roofs strong enough to support your weight. All too often you will find roofs that have leaked and deteriorated to the point at which they are unsafe to walk on. In this case, remove the roofing from the inside. If the attic has not been floored, lay some boards across the joists to provide firm support for the feet of your ladder. If both shingles and boards are in too poor a condition to be worth saving, remove both as a unit. Starting at the peak, insert the tip of a pry bar into the space between the highest board and the ridgeboard. Punch through the shingles so you can begin prying next to one of the rafters.

When you have an opening along the roof peak that has been left from the first board, begin removing the boards from underneath by using the crook of your bar as a hammer to raise the upper edge of each board enough to insert the bar and pry the board up. Once your opening is wide enough, you can progress rapidly, lifting the upper edge of each board, prying off the lower edge, and overturning each board so it forms a roll with the shingles still attached. The unit will work like the top of a rolltop desk. If the roof is too deteriorated to save, you can roll the entire roof down to the edge and load it into a truck to be hauled to the dump.

In some cases, the job will go even faster if you use a maul of about ten lb (4.5 kg) in weight. Use the top of the head as a poker to punch the boards next to the rafters so they fall free. Whether the maul or wrecking bar works better will depend on the condition of the boards and the way they were nailed.

Scaffolding and Security

Whenever you are working above ground level on walls, roofs or ceilings, you should consider using scaffolding. Small, light units are available that one person can handle. These provide sure footing and mobility that no ladder can offer. Especially when working on roofs, many people appreciate the firm, level footing just below the eaves. They do not need to grope over the edge of the roof to hit a ladder rung. Scaffolding provides greater safety; it cannot tip or slide sideways like a ladder.

The disadvantages of scaffolding are the greater difficulty in moving it from place to place, the cost of rental (unless you plan to use it enough to warrant buying a set for yourself), and the greater chance for a theft. A ladder can be easily taken home each day, so it is not sitting around where it can be stolen. Once you have removed doors and windows you cannot hope to keep thieves out of the building. Anything you cannot take with you at the end of the day should be heavily chained up. Buy a heavy chain long enough to fit over or around something sturdy, and lock up large ladders or scaffolding with a heavy-duty padlock. With these precautions, you are much more likely to find it when you return.

Where buildings are being torn down, theft is a common problem. Partially demolished buildings draw people like magnets. It may be a curiosity to see what really was inside a particular house; it may be an interest in watching the process of tearing down a building; or it may be the treasure hunter's drive to look for gold coins between the walls.

Most of these people are perfectly harmless, but there are too many looking for free handouts. If you approach someone tearing down a building, don't be surprised if the reception is cold; he will probably consider you a fraud seeking free materials.

The length that people will go to is amazing. One common practice is to show up as a relative of the original owner who promised you could have all the light fixtures, the china cupboard, or "some boards." These people often know exactly what the items are like either from prior knowledge of the house or from snooping about the night before.

While I was photographing a family project for this book, I watched a little old man drive up in a pickup truck. He took out a pry bar, walked up to the perspiring head of the household, and announced that the owner, whose name he knew, had said he could have "a few boards." He promptly turned around and began prying off some varnished, knotty-pine wainscotting. Needless to say, he was not thanked for his help when the family escorted him back to his truck.

With very old houses, the problem may involve people claiming authority from some historical society or city agency to salvage special fixtures, to supervise, or research, etc. I mention this only because I have encountered workers who have taken these fraudulent people at their word and watched sadly as antique, brass light fixtures walked away. If you have followed the procedure in the first chapter of this book, you will know what rights you own, and can send such people quickly packing. You will work very hard for the salvaged materials, so take every precaution to prevent freeloaders from sharing your profits.

Rafters

The rafters should be taken out next. Depending on whether the upper joists were floored over or not, you may have to lay some planks on the joists. Doing a balancing act on joists is tricky and dangerous. The plaster between will not likely support your weight if you take a misstep, resulting in a painful accident.

Start by removing the cross members—usually 2 by 6's (5.1 cm $\times$ 15.2 cm) or 2 by 8's (5.1 cm $\times$ 20.3 cm)—that connect the rafters on each side of the ridgeboard that runs down the ridge to join the rafters. Cross members, called collar beams, are usually spaced between every three or four rafters for added strength. It is fastest to use the large pry bar as a hammer, swinging the hooked end with the hook facing towards you. Strike the cross member where it attaches to the rafter. This should free it enough so you can insert the other end of the bar and pry it off. The other end will loosen if you force the free end away from the rafter. Sometimes you can push the free end far enough so the other end drops free. Usually, though, you will have to insert the bar, prying outward until the end drops free.

Continue to remove cross members until you reach the end of the ridge. Gather them up and make sure that they reach the ground safely. If you are low enough, toss them over the side. If not, carry them partway down the ladder before dropping them. As always, be sure no one is below you when pitching boards over the side.

The rafters can be removed in a similar fashion. Start at the eaves, where you begin with the first rafter in from either end. On smaller buildings, such as garages, the rafters may be toenailed to the top plate. For large buildings, including most solidly built houses, the ceiling joists are spaced the same as the rafters so the ends of both can be solidly nailed together. The entire joint unit is then toenailed to the top plate.

If the joists do not join the rafters on the top plate, you can hit the rafters loose from the plate and pry the joints apart with a pry bar.

If the construction is very strong, with joists and rafters heavily nailed together before being toenailed to the top plate, try starting at the ridgeboard. If you can pry the toenailed end loose there, you can use the length of the rafter as a pry bar to open up the joint at the top plate. Once you have opened the joint, a pry bar will finish the job easily.

Continue down to the end of the roof, and then do the other side the same way. As you remove each rafter, slide it over the edge of the roof, reaching down as far as you can towards the ground before dropping it. Unless they have been weather-damaged, rafters are an excellent source of salvage timber, so if you are too high up, carry them partway down before dropping them, to prevent them from cracking or splitting. Where they are notched to fit over the top plate, it is better to drop them so the other end strikes the ground to lessen the chances of breakage.

After taking out all the rafters, you will be left with two gable ends joined by the ridgeboard, with an occasional vent pipe or chimney protruding through the space left by the rafters. If the gable ends are not too large, you might consider leaving them intact. Many workers prefer to use them in the construction of garages or storage sheds without having to waste time taking them apart, only to reassemble them at the new construction site. These often have vents or windows that can be effortlessly added to your new building without your having to prepare rough openings or fitting windows.

If you plan to save the gable ends, start by removing the windows, following the procedure described earlier under "Windows." The gable end will now be joined only by spikes through the top plates. Loosen it by forcing the end of a pry bar between the plates and prying upwards. The hardest part is to prevent the gable end from falling to destruction when you pry it off.

Illus. 27. A rafter being pried off the top plate. Notice the notch cut out of its underside.

If the end is small, 10 ft (3 m) or under in width, you may have to bend the nailed joint inward and outward several times to loosen it enough so it can be forced over the edge. If the distance is very short, under 6 ft (1.8 m) from the gable peak to the ground, the fall will probably cause no damage, so pry off the last nails and let it drop. If the structure is large or the distance is much more than 6 ft (1.8 m), fasten a pair of ropes to the bottom of the gable and secure them over the roof into a window or to the chimney in order to ease the fall.

To secure a rope to a window opening use a length of 2 by 4 (5.1 cm × 10.2 cm) longer than the span of the window opening. Lay it crossways on the inside of the opening, and use a spike on each end to hold it to the wall. This provides something to tie support ropes to when lowering heavy objects.

Illus. 28. Removing a gable end intact and ready to use on a new garage.

Once you have lowered the end safely to the ground, you will probably need either a helper or a small winch to load it on a trailer or into a pickup bed. Be sure to follow the safety regulations for width on the roads you travel. You may have to stand it on end so it is not too wide to haul.

The techniques just described cover a gable roof with no protrusions. The process will be more complicated with the many common architectural variations. Dormers are among the most common additions to a straight-gable roof. There are more windows to take out, some of them very decorative. The shingles will come off the same as those on the rest of the roof, except that you will have to stop and change angles. Beneath the shingles are very short roof boards covering short rafters that form a miniature roof, which a dormer really is. Pry the boards and rafters off just as you did for the rest of the roof.

There is also a small amount of siding to deal with on three sides of the dormer. This should be removed after you have removed the roof boards. Start at the upper back edge of the dormer in the angle where the boards joined the roof boards. Insert the pry bar between the board and the framing, and pry it off. Follow down the side, prying off the other boards. The front siding should come off the same way if you have removed the strip around the window.

In most cases all that is now left of the dormer is upright 2 by 4 (5.1 cm $\times$ 10.2 cm) studding and a layer of plaster. Pry out the 2 by 4 (5.1 cm $\times$ 10.2 cm) and smash out any remaining plaster and lath with the pry bar.

Flat Roofs

Flat roofs are also common. These have been constructed so the ceiling joists play the part of rafters, with the roofing material applied right over them. They are not as strong in design as the triangular design of gable roofs. To counterbalance this, they are usually built more sturdily, either by spacing the joists closer together, using heavier joist stock, or by additional reinforcement.

The common practice is to provide an overhang on all four sides that keeps run-off water away from the building and offers better protection for the upper walls and windows. Begin removing a flat roof at the eaves by removing the band that encloses the ends of all the rafters. Start at one corner and pound outward at the joints with the side of a pry bar or a hammer, and then pry the band off at all the joints. Where the roof overhangs, you will find a series of short rafters called lookout rafters running at 90° to the main rafters. These are easy to pry out once the band no longer holds them by the outer ends.

The remaining rafters will be toenailed to the top plate on one end and either to the opposite top plate or to a support beam, depending on the size of the building. If the building is large, requiring a support beam, you will find a joint where the two sections of rafter are joined on the beam.

The sections may be overlapped and nailed through the face, or butt-joined with splicing material face-nailed to one or even both sides.

The joints may not be easy to disassemble, so first you may want to pry the unit off the beam. Then pry off the opposite end where it meets the top plate. Take apart the joints later. In some cases, you may want to preserve the joints so the full length of the rafters can be used on some other project. If you decide to separate the rafters at the joint, you may find that some can be pried apart with just the bar. For stubborn ones, add a hammer and wedge. For those butt-joined by overlapping plates, try using an old axe to split the plates so they can be broken off, exposing the nails for the pry bar.

Hip Roofs

Hip roofs are easily recognized by the inward tilted pyramid you will see from the end view. This style has become extremely popular, although its construction is slightly more complex than a straight gable. Their complexity does not, however, cause as much trouble in dismantling as sawing all the unusual angles caused the builder.

The hip roof has two hip rafters that run from the ridge board at an angle down to the outer corners of the roof. From the top plate on both ends and sides, short rafters called jack rafters connect the hip rafters to the top plate. The only difference between hip roofs and gable roofs is removing the many jack rafters. Their outer ends are either toenailed to the top plate or face-nailed to ceiling joists. Remove this end as described earlier. The upper ends are toenailed to the hip rafters at a double angle. Because the lower side is attached at an angle, no nails can be driven from that side. All you have to do to loosen the joint is to strike the joint from the lower end. You may find that a 2-to-5-lb (.9-to-2.3-kg) hammer is the best tool for the job. You can also pry off the outer end and twist the jack rafters in the direction of the nailheads. Try both to see which works better for you.

Gambrel Roofs

Gambrels, used most often for barns, are among the most complex designs for houses. Some people have built gambrel houses just because they like the design, but they do offer an advantage over gable styles by providing more usable headroom and space in the attic. As a result, you are more likely to find a finished attic with gambrels than you are with other designs. The roof and rafters can still be removed the same way from the outside. Think of a gambrel as two separate roofs, one with a steep slope and one above it with less slope than most gable roofs would have.

Because of the extreme angle, you will need scaffolding to work on the lower slope when removing shingles and roof boards. The upper slope is gentle enough, so you can walk on it safely without supports. Between the two, a horizontal timber called a purlin connects the two sets of rafters. The big difference in gambrel construction is in the joints where the lower rafters and upper rafters meet the purlin. The rafters are notched to enclose the purlin, resulting in a strong construction. Because of the angle of the notch, the rafters must be removed first at the top plate for lower rafters, and at the ridgeboard for upper rafters. Then by lifting or twisting them sideways, the purlin joints can be freed. Remove all the rafters first.

Below the upper rafters, you will find a series of collar beams that span the purlins. These serve as ceiling joists if the attic is unfinished or, if the room is finished, these will serve as both floor joists and as ceiling joists. Pry out the collar beams and, from the bottom ends, the lower rafters, twisting them sideways to free them from the purlin. Then remove the purlins and any remaining upright supports. From this point on, the building is just like any other.

The styles of architecture have countless modifications, but with directions for gables, gambrels, flat roofs, dormers, and hip roofs you will find that the many variations are really only combinations of these types, so you should be able to tackle any design successfully.

Porches

If the building has porches with overhangs, treat these as just what they are: miniature houses. They have a small roof, walls, and a floor, just like any others in the building and, unless they were later additions, were probably built with the same materials. Take these apart from the top down while you are doing the rest of the house.

If the porch balcony is enclosed by a railing, remove the supports below the railing. Sometimes these are ornate enough to preserve for other projects. If they are tightly fitted between the railing and the balcony floor, use a small jack to raise the railing enough so you can pry out the supports. Use any kind of jack, including a car jack. Fit in a length of wood between the jack and the underside of the railing to increase the effective length of the jack if it is too short. After you have removed the supports, the railing can be pried away from the building with a pry bar. You may have to disassemble the sections of railing before transporting it.

Some porches are supported by large columns. While these appear heavy, they are generally light enough to be lifted with one hand. They are hollow, made from thin wood strips nailed together. Even if you have no plans to include them as part of your next design, save them. Sawn into short lengths, they make excellent ornate birdhouses with the addition of a roof, floor, and entry hole.

Illus. 29. Pry out the support columns to remove a porch railing. For large porches, a jack can help.

Attic Floors

If the attic room was finished, you will be dealing with a finished ceiling, walls, and floor one level earlier than you will with an unfinished attic. Let us assume that the upper level has a finished floor. The upper layer of flooring will already have been removed if it was usable carpeting or hardwood. The layer beneath is called the subflooring. This is usually shiplap, having a rectangular notch cut from opposite sides of each edge so the boards can overlap about ¾ in (19.1 mm). The overlap helps prevent the warping and squeaking that are so common with boards that are only butted together.

To remove shiplap, start from the side where the first board was installed. Otherwise, you will be prying up against one of the overlaps, causing broken boards and much more work for yourself. You can quickly tell which side the first board was laid on by examining and comparing the edges of the boards on each side of the room. The last boards to be installed have been sawed or planed to fit against the wall, so you should start with the other side.

Removing flooring is often tedious, and sometimes difficult. You have no chance to pound floorboards off from below, as you may choose to do with roof boards. You will, instead, have three options. First, you can use a pry bar, preferably one with a sharply ground tip. By repeatedly forcing the tip between the joist and the edge of the board, you can raise the board at each joist, where it will be held by two or more nails. If you start at one tip, the board will be easier to raise along its length.

Even working carefully, though, you will find that many boards split because pressure is being applied to only a small part of the board at one time. As another approach, use a nail puller. Set the open "beak" of the puller around the head of each nail, and drive the handle down. This action forces the beak into the wood below the nailhead. When you tilt the puller to the side, a short lever presses against the board, tightening the beak on the nail. Further pressing will lift the nail out. Actually, it takes much longer to describe the action than to pull out a nail. Once you have pulled out a dozen or so you will develop a smooth and efficient technique.

A nail puller may seem to take longer than a pry bar, but you can save every board with no danger of splitting, and remove all the nails at the same time. When you consider the time you would spend hammering out the nails, and the easier way of stacking or loading the boards for transport, the nail puller will sound better still.

A third method is one I have seen used by a few experts who have torn down many buildings. After wrestling with a floor or two, they have concluded that both pry bars and nail pullers are backbreaking, as well as slow. Their solution has been to build a special tool just to remove flooring. The tool works equally well for roof boards.

The best tool I have seen is one designed by Don Arndt of Polar, Wisconsin. Using an acetylene torch, he cut a 2¼-in (5.7-cm) "U" from a ¾-in (19.1-mm) steel plate. At the bottom of the "U", he welded a socket to hold a 4-ft (1.2-m) section of 2-in (5.1-cm) steel pipe. The two prongs of the "U" slide under the floorboard on each side of the joist. When backward pressure was applied, the back of the cutout used the joist as a fulcrum to lift each joint up with very little effort.

For efficient removal of floorboards, this method is unbeatable. If you plan to do more than one building, you might consider building an Arndt flooring puller. It will definitely increase your speed and efficiency, but you will have to decide if it is worth the trouble of building one or having one built for you.

Illus. 30 (above left). Using a nail puller to remove flooring takes longer than using a pry bar, but prevents splitting and removes nails at the same time. Illus. 31 (above right). Close-up of Arndt flooring puller. Illus. 32 (below). Another type of flooring puller welded from separate pieces of steel.

Earlier I recommended that any valuable wall or floor coverings be removed before the real work begins, so anything left at this point should be damaged or low-grade material. If you have not already done so, remove the mouldings at the base of the wall. This may be only a wide base shoe, or it may be a base shoe and a quarter round that extends out from the base shoe. You may also encounter special moulding designs, but all of them should be face-nailed to the wall. Use an old chisel or fine-edged pry bar to pry off the mouldings; start from one corner and continue around the room. You are now ready to start on the walls.

Old panelling on the walls may involve vertical corner mouldings, so pry these out the same as you did for baseboard mouldings. Then slide a bar between the end of the panelling and the wall to pry it off.

Beneath it, you will be dealing with plaster and lath, or Sheetrock™ and dry wall. The techniques needed for efficient removal of each differ slightly.

To remove Sheetrock™, start with a hammer, pounding a hole high up in the space between two studs. Insert a length of sturdy pipe or heavy steel rod about 4 to 6 ft (1.2 m to 1.8 m) long—such as an old ice spud—into the hole and slide it down inside the wall so only a short section is left for a handle. Then pull the handle towards you, opening the wall as you would with a giant can opener.

One of the special pry bars mentioned in the tool section can also be useful. With only a few quick swings of its flattish head, you can use this pry bar built from two separate pieces of steel to break a channel from top to bottom, and then use one of the hooked ends to pull the sections inward. The heavier weight of this bar gives more power to each inward pull. This leaves two plaster ends that can be pulled inward with either the hook end of a pry bar or sometimes your gloved hand. Tear the material off at the studs on both sides and go on to the next set of studs. This goes much faster than removing flooring. An average room can be finished in about a half-hour or 45 minutes.

The older plaster-and-lath construction is sturdier, requiring more effort to remove. The can-opener technique will never work on wooden lath. Here again, the flat head of the pry bar will be of use. If you swing at the plaster, the lath inside may only flex, breaking the plaster but leaving the lath intact. Often, you can punch the tip of the bar between the lath and loosened plaster to break it out. This is usually the easiest way, and it allows you to save the lath for other projects.

The lath can be used to build rabbit hutches or other small animal cages. I also know of one worker who strings the lath together with wire to build snow fences. The broken ones, or splinters, become handy fire starters, so even these should be saved. If you don't want them yourself, usually word of mouth is enough to get somebody interested in them. As more people heat with wood, lath will become even easier to give away, or perhaps sell. Use binder twine to tie the better ones in bundles so they do not scatter.

Illus. 33. To remove Sheetrock™, first punch a hole in the wall with a chisel, then open the wall with a long metal rod.

Illus. 34. Old lath can be used for snow fences and rabbit hutches. Tie the lath into bundles to keep it out of the way.

Upper Ceilings

When you have the highest walls taken down to the top floor, and have removed the upper flooring, take out the upper ceiling. Remove it from above, striking downwards between the ceiling joists with the curved end of a pry bar. From above, you will not only find the job easier, but you will be away from the dust that will fall into the room below. Since you will be left with only joists to stand on, set some boards across the joists as you work. It is easy to lose your balance doing a tightrope act on the joists.

Whenever you remove interior plaster, lath, or Sheetrock™, the materials will form a large pile of rubble. Too often, I have seen workers push everything down a stairway, moving aside just enough below so they could climb over the rubble to get up and down the stairs. This is asking for trouble. You can trip on the rubble and land on nail points all the way down the stairs.

It is better to stop periodically to remove rubble before it builds up to dangerous mounds. One suggestion is to open sections of flooring and ceiling in all floors so debris can be pushed through the holes all the way to the basement, where it will be completely out of the way. At least on the lower floor you should be able to push all the debris into the basement, if the building has one. Brooms and shovels are of little use for removing the debris. Instead, take a length of 2 by 4 (5.1 cm × 10.2 cm) and nail a 2- or 3-ft (.6- or .9-m) section of board to the wider side of one end. Use just one nail right through the center. The unit will now act as a pusher for all debris, so you can maintain clean, safe floor areas.

With the upper ceiling and flooring gone, the ceiling joists are easy to remove. Pry them loose from the top plate at one end, and move them sideways to loosen the other end. Slide them endwise over the side, lowering each one as much as you can before dropping it. The joists may run in several directions, some being supported by one or more main beams. The joists will be toenailed to any partition they cross, as well as to the plate on both ends. Sometimes the ends of the joists may be notched to fit around a beam, but all the joints should pry loose with little difficulty. Wherever a chimney passes through the joists, you will find headers—short pieces of joist lumber usually doubled for extra strength—running at 90° to the joists. The joists that are normally doubled on the other sides of the chimney are called trimmers. You will have to pry both sets of doubled lumber apart while removing them, but you will be gaining even more choice lumber.

If large main beams have been used to join sections of joist, you may need help to remove them. Some of these are heavier than they look, so use care when you pry them out. Use a rope-and-pulley system to ease really large beams down to prevent any chance of damage, since old beams can add fine touches in redecorated buildings.

Illus. 35 (above). Ceiling joists with the ceiling plaster and upper flooring removed. A pry bar swung against the ceiling joists from above broke the bridging loose between them. Illus. 36 (below left). Prying out a ceiling joist with a pry bar. Notice how the joists are spliced in the middle. Illus. 37 (below right). With one end of the joists pried loose and set on the floor, you can work on the other end.

Exterior Walls

The method you use to dismantle the exterior walls depends on the type and condition of the siding and sheeting. The materials can vary even more than materials used for interior walls. You may find nearly decomposed asphalt, painted wallboard, or even aluminum siding that was applied over previous siding. Decide whether the siding is worth saving, and deal with that first. Artificial asphalt—brick—is one of the more common siding materials. If it is in good condition and you work on a warm day while prying it apart from the top down, you can usually save nearly every piece. This material is no longer popular for houses, but it can be used to cover a garage or storage shed. It is maintenance-free, and lasts surprisingly well.

Illus. 38. If you plan to remove sections of wall intact, remove any remaining nails from the outer siding. Otherwise, you will destroy your saw blade.

If you decide not to save the exterior asphalt, still remove it before working on the sheeting below. Work from a ladder outside, using a pry bar to pry the asphalt loose at nails where it hangs up, and your gloved hands to tear it away in sheets. The material comes off quickly.

Exterior board siding can vary greatly in quality, condition, and construction. The boards may be nailed either to a sheeting beneath them or directly to the studs. They may be applied horizontally and overlapped, or they may be of vertical tongue-and-groove construction. They may even be board-and-batten, consisting of wide vertical boards with their joints covered by narrow strips nailed to the studs, or to horizontal 2 by 4's (5.1 cm × 10.2 cm) between the studs.

If you plan to save the siding, work carefully, because the boards may be brittle from many years of drying. For overlapping horizontal boards, start at the top and work down with a thin pry bar. For tongue-and-groove boards, start at the bottom and work up, just as they were applied, to prevent splitting the edges. For board-and-batten construction, pry off the battens and pry out the boards, which are seldom held by anything except the battens and one nail in the center.

If the boards are nailed directly to the studs without undersheeting, try to tap them off from the inside. Experiment with the pry bar as a hammer, hitting the boards next to the studs where they are nailed. Sometimes this works very well, but if you start splitting boards, try prying instead. Sometimes the only way to save boards with any consistency is to use a nail puller.

Another type of siding common in older buildings is brick veneer. People are often puzzled by this term when they see some of the brick that has been removed. "Isn't it real brick?" they ask.

The brick is real, but the construction is artificial. A brick wall is constructed of overlapping bricks that tie the wall together to provide great strength. Brick veneer consists of only one row of bricks. Instead of locking together in thickness with other bricks, they are only set against the wooden sheeting. The joint offers no strength at all. The only attachment between the brick and wood is a large number of long nails driven at intervals into the boards. These are called brick ties, as they offer support for the mortar joints to hold the bricks against the wood.

As for taking them apart, it is easier than you might have guessed from the description. All you have to do is start an opening at the top, and you can peel off the whole wall with your hands or a pry bar, whichever you prefer. The best thing about brick veneer is the large number of bricks you can salvage for other uses with only a small amount of work. To reuse the exterior sheeting, though, you will spend some time pulling out the brick ties.

When the exterior siding is off, you may have only bare studding or one or two layers of sheeting. It is not unusual on older buildings to find two exterior layers of 1-in (2.5-cm) sheeting, their joints staggered and a layer of heavy tar paper sandwiched between the two layers. With a single layer of exterior sheeting, most workers start from inside, hammering one edge of the sheeting with the curved end of a pry bar so the other end of the bar will slide into the joint for good leverage. With a double layer of exterior sheeting, the best solution is to take the wall sections apart first. Then, with the sections flat on the ground, place a brick or other prop under the top plate to create a space below the sheeting. Using a heavy hammer or the curved end of a pry bar, pound the sheeting off in the space below. This goes quickly with two workers, one on each end of the boards, pounding in unison.

Illus. 39 (left). This looks like a normal brick wall, but it is really brick veneer.
Illus. 40 (right). Beneath brick veneer, you will find a series of brick ties. You can peel the bricks off the wall beneath with your hands.

Large public buildings such as schools and churches often were built with double sheeting, one layer on each side of the studs. The widths of the boards were different, so the joints would be staggered, both for slightly greater strength and freedom from air currents. The best way to remove the boards is to do both the outside and inside at the same time from top to bottom. Work from inside on scaffolding or from a steady ladder. Direct a blow at the upper edge of both inside and outside trim before reversing the bar to pry them off in the space created by the blows.

Illus. 41. Public buildings were often sheeted on both the inside and outside. Use a pry bar to remove both layers at once, so you will not have to move the ladder or scaffolding too often.

Another idea worth considering is to remove the exterior siding and interior wall covering, but preserve the studding and exterior sheeting just as it is. If you plan to build a garage or storage shed, or know someone who might be planning to, think of the work you can save by keeping the construction intact.

To do this, use an old chain-saw blade to cut through the sheeting, top plate, and all supporting timber all the way to the floor. Make the cut on one side of the stud nearest to an 8-ft (2.4-m) length. Two workers can usually pick up the cut sections and load them on a trailer or pickup truck. When you assemble your new building with these "prefab" units, add another stud as a nailing surface for those portions without a stud at the end. You will also have to run short segments of 2 by 4's (5.1 cm $\times$ 10.2 cm) across the cut top plate and sill to splice them together. When you consider the time spent taking the structure apart and reassembling the pieces, the technique sounds more appealing. If you add the intact gable ends discussed earlier, you will come close to a preassembled building.

Stairways can be removed in sections as soon as you have no further use for them. Several patterns are used to build stairs, but they all consist of long, heavy timbers called stringers, and short boards that form the treads and—in most cases—risers that fill the spaces in front between the treads.

Illus. 42. Use a chain saw to cut through the sheeting next to a stud.

Illus. 43. By cutting through the sheeting, sections of walls or even roofs can be removed intact and ready to install on a new building.

While the treads are intact, remove handrails and any fancy ornamentation that might be involved, such as turned balusters or posts. Sometimes these can be fancy enough to interest antique dealers. Most pry out easily, though some are bolted or screwed on. Save all the parts, especially spindles or turned top knobs.

Start removing the treads at the top. Tap upwards on any overhanging edge to loosen a tread. Occasionally, the treads may fit flush with no overhangs, so it may be easier to remove the riser first so you can get at the tread. Remove the treads and risers all the way to the bottom, exposing the stringers—either two or three, depending on the quality of construction.

Illus. 44. Stair risers are always in demand for new construction.

With the treads and risers and rails gone, work from a ladder to remove the stringers. These may be supported by additional timber beneath, so pry these out and then work on the stringers. Try a pry bar along the joint between the stringer and the wall. Stair stringers are usually installed firmly, so you may have to use a hammer and wedges to remove them. They are time-consuming to build, so save them for new construction—where they will be welcome if they are in good condition.

You should be left with only the walls and either bare studs or exterior sheeting you plan to remove after those walls are lowered. The next process is to disassemble the interior walls. Remove the top plates first. Then you will have a line of studs toenailed only at the bottom. These are easy to pry out. Where electric cables pass through holes in the studs, pull them back through the holes out of your way. Where they hang up or are attached, cut them off with a hacksaw or an old axe.

Illus. 45. Whenever flexible conduit is in the way, pull it back through the holes in the studs or joists.

Illus. 46. Chop off the sections of conduit where they cannot be pulled through. Use either an old axe or a hammer and cold chisel.

Heating ducts may run anywhere inside the studding. These are light sheet-metal, nailed to studs or to trimmers on each side. As the studs are pulled away, the nails will let go; the heating ducts can then be lifted out. These can nearly always be used again.

Illus. 47. Removing a heating duct from between two studs, which are lightly nailed into place.

Illus. 48. After nails have been loosened, the duct lifts out, and is ready to install again.

Plumbing, too, runs through the spaces between studs. Some of this can be reused directly if you can turn the fittings with a pair of pipe wrenches. The large, iron soil-pipes, though, are nearly impossible to save. The joints are assembled with molten lead, so the effort of trying to get them apart is wasted time. Save all such parts for sale as scrap metal. They can be broken with a heavy hammer or, if necessary, a cold chisel.

The perimeter walls of studding are easy to disassemble at their corners. In most cases, they were built flat on the floor before being raised into position. Begin prying at the corner to free the top plates, which overlap to hold the top together. Lengthy reinforcing timbers may also have to be removed, to allow the walls to separate. A section of studding or joist used as a pry may help. Keep prying any last nails so the corners are free. Then you should be able to push the walls outward so they will fall. Once they are down, you can quickly remove any sections of sheeting that for one reason or another were inconvenient to remove before. Follow by pounding off the top plate and removing all the studs.

Illus. 49. Some walls are easier to take apart after they are on the ground. Loosen the corners and pry the base loose with the pry bar, followed by a section of joist for greater leverage.

Illus. 50. Some workers prefer to collapse walls first before removing the sill plate or studding.

Where the windows were fitted you will have extra wood to take out. Remove the studs on each side, prying them off the header at the top of the window opening, and the double sill at the bottom. The headers and sills will be easy to pry out after you remove the short trimmers that were fastened to the studs. Two or more 2 by 4's (5.1 cm $\times$ 10.2 cm) are toe-nailed between the studs at top and bottom. Pry those out to complete the window-opening disassembly.

Most older buildings were built with hollow walls and no thought about insulation. If you do find insulation between the walls, it will most likely be the blown-in type that has been added later. If you can prevent it from being rain-soaked, you can save the insulation and use it again. Gather it up in plastic lawn and leaf bags to store until you need it. A few old buildings were designed with insulation in mind, notably locker plants, icehouses, or meat-packing plants. If you have the chance to tear one of these down, you can plan on a large supply of insulation. At today's prices, you could save a great deal of money on insulating your own house, compared to the cost of hiring a commercial company to do the job.

The studding provides a large quantity of building material, so pile it well out of the way to prevent it from being damaged when you topple the chimney. Now that all the walls are gone, there is little for it to damage as it falls. If it is small, usually a few blows from a heavy maul at the base bricks will weaken the joints enough so one person can push it over the way it should fall. For larger chimneys, you may need a four-wheel-drive vehicle with a long chain or cable to pull it over.

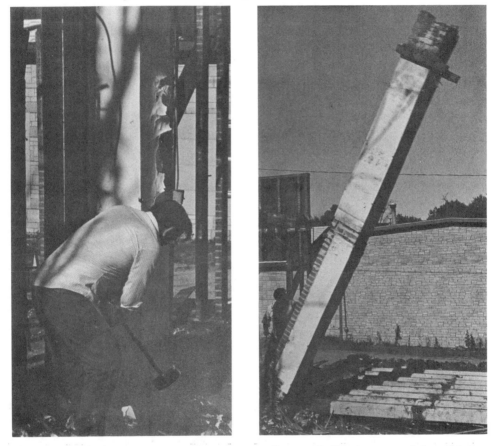

Illus. 51 (left). As soon as you have removed enough flooring material and joists so nothing will be damaged, knock down the chimney. Weaken the base with a maul. Illus. 52 (right). Push the chimney over. The chimney will break into sections, but the brick can still be salvaged.

If the chimney is sturdy concrete block, which may be the case if the chimney was added in recent years, you may want to dismantle the chimney while it is standing to prevent damage to the blocks and flue liner. If you decide to do this, take it apart in sections as you take the building apart. Do not be tempted to climb a long ladder leaning against an unsupported chimney. Any joint might break, toppling both you and the chimney. So, if you did not disassemble the chimney earlier, topple it now.

If the chimney is very strongly built, or if there are heavy portions left inside, such as the opening for a fireplace, you may need more power to topple the chimney. A cable high up around the chimney with tension applied by a pickup tractor or four-wheel-drive vehicle will usually topple even sturdy chimneys. You may need to weaken the chimney just above the fireplace so it will fall. Some parts may be damaged, but you can still save most of the blocks or bricks. As you chisel them apart with a cold chisel and hammer, salvage any unbroken sections of flue liner, if there is one; older chimneys did not always contain flue liners, depending instead on tightly tied bricks to guard against fires.

Illus. 53. A hammer and cold chisel will break the mortar joints in a chimney.

If portions of the fireplace remain, salvage any metal parts. Any dumps, dampers, draft controls, or metal fireboxes should be removed by chiselling away brick or stone to get at them. Most of these parts can be used again. When you have them all out, use a heavy maul or hammer and cold chisel to knock down the rest of the brick or stone so you can get at all the flooring.

The cast-iron vent pipe and any other piping should now be pried or pounded out, leaving only the level floor and the supports beneath it.

Pry out the flooring and any layers of subflooring beneath, using either the pry bar or the flooring-removal tool described earlier. The joists will then be revealed, ready for removal.

Illus. 54. Pry out any support boards or headers around the soil pipes.

Removing Floor Joists

As you did while working on the ceiling joists, slide several boards on top of the joists to provide good footing. You may feel comfortable working on the ground floor, but if your foot slips between the joists you will suffer severe bruises or lacerations when your thigh or groin strikes the joist. Floor joists rest on wood girders or beams that support their middle area, and on sills that support their ends. Sills are usually 2 by 6 (5.1 cm × 15.2 cm), although in older buildings the size may vary. The ends of the joists may be butted against a header joist that stands on edge to enclose the open joist-ends. Where the joist-ends butt against the header joist they will be toenailed to the header.

Illus. 55. Balancing on a beam or remaining joists can be tricky. Place some lumber or flooring across the other side to walk on.

If the joist is too short to reach to the opposite header joist and sill, there will be some kind of splice on the girder or beam that runs down the middle. The more recent practice has been to overlap the two sections of joist on top of the beam by face-nailing the sections together. The spaces between joists will be filled on top of the girder by short sections of joist material called bridging; this material will be toenailed to the joist at each end.

To remove joists attached in this way, first remove the bridging. The way it is fitted allows no room to insert a pry bar, so use a maul to pound them out by striking them close to each end. Then work along the opposite ends, removing the header joists with the maul and pry bar. Once they are exposed, the ends are usually free because they are normally only nailed to the header.

To gain extra headroom in the basement, the girder may have a ledger piece nailed low on the face-side to support the end of the joist. The end of the joist will be cut out in an "L" to fit around the top and side of the girder. The joists are both toenailed to the girder and face-nailed to each other. The easiest way to remove joists fastened this way is to free the opposite end so you can use the length of the joist as a pry bar to loosen the joist from the girder. If the thin section of joist on top of the girder breaks, the damage is of no importance, since you would normally be sawing the extension off before the joists could be used again.

Illus. 56. Floor joists are often easiest to remove by loosening one end and walking sideways to twist the other end free.

A last method of construction common in older buildings is the use of notched beams. Notches were cut out of the beams to allow the ends of the joists to slip in. Since the joists could go nowhere but up, and their tendency to leap into the air with the house on top of them was slight, these joists were seldom nailed. It may seem surprising to lift them all out without doing any prying, but log cabins stayed together without nails— and the notched-and-fitted construction is similar. So just lift out the joists and be glad some early workman did a careful job of fitting that speeds up your own work.

Illus. 57. This beam has been notched to allow joists to slip inside it. Where joists are fitted like this, you will seldom find any nails.

Joist removal should proceed smoothly unless some joists pass through a poured concrete or block support. This may be the case with some additions because the entire structure was not planned in advance. If the joists are embedded in concrete, the only solution is to saw them off on both sides of the concrete. An old chain saw will speed up the job, but be careful that uneven cement does not protrude enough to cause the saw blade to kick out. When all the joists have been cut you will have one free end on each joist, but they will be too short to use as normal joists on your next construction project. You will have to find other uses for them.

Floor Beams

Beneath the joists, you may encounter three different means of support. You may find solid wood beams, usually hardwood such as elm or oak, steel beams in the shape of an "I", or girders built up from many layers of 2 by 6's (5.1 cm × 15.2 cm) or 2 by 8's (5.1 cm × 20.3 cm).

Solid hardwood beams were used for buildings many years ago. They are strong, but too thick to dry effectively in a short time. As a result, they often developed severe twists or warps after a few years. You may find whole floors sagging because of warped beams.

The larger beams may be too much for one person to handle. The beams were generally cut from any available wood, so they might be either hardwood or softwood. Pine or fir beams hold a higher moisture percentage than hardwoods. As they dry out over the years they lose weight, so usually even large beams can be handled by one person.

If the beams are solid elm or oak, however, they will be incredibly heavy. If they have developed severe warps or twists, you might decide

88

to saw them into shorter lengths. They can still be used for decorations on the other projects mentioned in the last part of this book. The shorter lengths will also be much easier to handle.

Illus. 58. Most houses have girders built up from many layers of 2-in (5.1-cm) lumber. Drive a pry bar between them with a hammer. After you have separated them slightly, complete the prying with two pry bars or an old ice chisel.

Steel beams and hardwood beams straight enough to be valuable in one piece will take extra muscle power to remove, load, and unload. Rent a small winch that can move any normal beams. Fasten the cable around one end of the beam, and chain the other end of the winch to a tree or the frame of a vehicle. If portions of the foundation remain to wall in the end, use a maul to break an opening so the winch can pull straight endwise on the log.

Be sure everyone, including yourself, is well out of the way as you operate the winch. When the other end of the beam jumps off the foundation, it will fall into the crawl space or basement, jerking violently as it does so. If the basement is deep and narrow, the beam may end up on an extreme angle, but once it has fallen the winch can haul it up the side of the foundation until it again lies flat.

When you have moved it within a few feet of the truck, build an inclined plane from sturdy boards leaned against the tailgate. Then you can winch the beams into the back of the truck for transporting.

With the beams out, salvage any last materials from the basement. If the beams are supported by heavy steel pipes, known as Lally™ columns, collect these for reuse or for scrap metal, depending on their condition. Save any last sections of metal piping, copper tubing, etc., to sell as scrap.

Special Considerations for Barns

Barns are usually far more dangerous to tear down than houses. When a house has deteriorated to the point of uselessness, the owner normally tears it down to build something a person can live in. Barns, however, are often left to deteriorate until the owner decides they are too dangerous to leave around. Barns may be built with heavier beams than houses, but the spacing of rafters and the large size spread their weight over such a wide area that the whole structure tends to bulge outward in all directions. I have seen some barns that are unsafe to go near, let alone try to salvage. If you are given an opportunity to tear down one of these sprawling, sagging monstrosities, I would encourage you to pass up the chance, leaving it instead for someone who places less value on his life.

Barns may be torn down because the foundation has deteriorated or because they stand in the way of other progress. If so, the structure of the barn may be rock solid and as safe as any house, as long as you remember that all the parts are larger, so any part that falls will be that much more dangerous.

Assuming the barn you plan to tear down is in reasonably sound condition, most of the project can be completed without complicated machinery. The following tools will be needed: nail puller, pry bar, claw hammer, wrenches, chain saw (a bow saw may work), ladders, hacksaw, 8- to 12-lb (3.6- to 5.4-kg) maul, long pieces of ¾-in (19.1-mm) rope or chain, four-wheel-drive vehicle or tractor, and a pair of leather-faced gloves with high tops, preferably reaching to the elbow. Thick-soled shoes are essential because the risk of splinters and punctures can be high unless proper precautions are observed.

Most people who plan to tear down a barn already own most of the necessary tools. The one tool that might be unfamiliar is the nail puller.

"Do you have a nail puller?" a relative asked me when he learned that I planned to tear down a barn.

"A nail puller?" I repeated, shaking my head.

"They'll save you a lot of wood," he answered, handing me the unfamiliar tool.

During my work on that barn I grew to respect this ingenious cross between a hammer and a pry bar. Although mine was stamped 1905—making it two years older than the barn it was helping to tear down—nail pullers can still be found at any quality hardware store. They cost about $15, and return many times that amount in wood saved during one day's use. In seconds, the narrow "beak" is inserted below the head of the nail where it grabs the shaft. Then the nail can be removed effortlessly, leaving the board unsplit and nail-free.

After the tools are collected and a written agreement between the barn's owner and you has been secured, the real work of dismantling the barn begins. Before pulling that first nail, pause a few minutes to appreciate the weathered structure. It has been standing for a very long time. Often, the structure is as solid as the day it was built, but for matters of convenience—a housing development or an enlarged yard—the barn must be torn apart.

An engraved cornerstone will usually reveal the barn's exact age, and onlookers will recall the highlights of its history, but you will discover on your own many important features about the barn. Perhaps one of the most exciting discoveries about an old structure is the date of the barn-raising pencilled on a main beam. Because the building of a new barn indicated prosperity, it was an extremely important event to our ancestors, and they recorded it even more carefully than they did the building of their homes. Not only was the date inscribed on a main beam and engraved on a cornerstone, but the event was probably mentioned in the local newspaper as well.

Salvage Items

Interesting and often valuable items to be salvaged include wooden pulleys, hand-forged latches, heavy metal hinges, weather vanes, lightning rods, rustic milking stools and barn-wood gates. These should all be removed before the actual dismantling begins. If the barn is over 75 years old, most of the metal work will be hand-forged rather than machine-stamped. This means each of these is a one-of-a-kind item with special charm. Be sure to check for pulleys at the upper end of the roof beams. Used for raising hay bales into the loft, these early pulleys with hand-forged metal parts are valuable antiques.

Barn doors are always a conspicuous part of a barn's exterior. The sliding door is typical, with metal fittings on the door riding on metal rails. Sometimes the rails have been bolted on, and the bolt ends flattened against the nuts. The same technique is often seen on lock or handle bolts, or on the metal plates that fasten the door into the rails. If you run into this situation, use a cold chisel to open a space between the wood and the nut. Then saw off the bolts inside the nut with a hacksaw. For metal parts assembled with screws, remove the screws with any large screwdriver. Even if the screws have rusted, the wood has usually shrunk enough so the screws will turn.

Illus. 59. Sometimes you will receive unexpected help while taking the hinges off a barn gate.

Large nails, too, have often been used to hold rails and metal parts on barn doors. The points will be bent over and driven into the wood on the inside. These are easily pried up with a cold chisel driven under the points. You will sometimes find nails that have been beat into a "C" shape before pounding the points into the wood. This was safer than leaving points where they could stab someone, but it is no help for the one removing the nails. The nail puller is the best tool here. Punch the beak in on each side of the highest part of the curve, and pull the point out by its middle.

Straighten the nail's point slightly so the pry bar can pull it out from the other side when you pry against the metal part.

Be careful when you reach the end of the rail fasteners holding the door in place. Barn doors may be built from dried softwood, but they are deceptively heavy. If the door falls forward on anyone, severe injuries will result. So clear the area of any bystanders before you remove the bolts or screws. Also watch where your ladder is placed if you are working high up. Anticipate where the door will fall, and be sure neither you nor your ladder will be in the way. When the door finally falls, it will raise a real dust cloud.

With the door down, you might as well take it apart so it is out of the way. Normally, you will have no use for any door that large, and transporting it intact would be difficult. The door will be nailed through the front into cross boards on the back. If the nails are driven through and bent over as usual, pry up the nail points so the crosspieces can be pried off the back; the door will be reduced to a pile of boards you can stack or haul away.

Also valuable are the weather vanes and lightning rods. These are easily pried off the roof ridge, but first you have to reach the ridge. The height can be terrifying to anyone unused to it. If you are bothered by heights, be sure to employ a worker who will be able to salvage items from the roof. Save all the salvaged items, even if they are slightly damaged or have missing parts. They can often be restored with little effort.

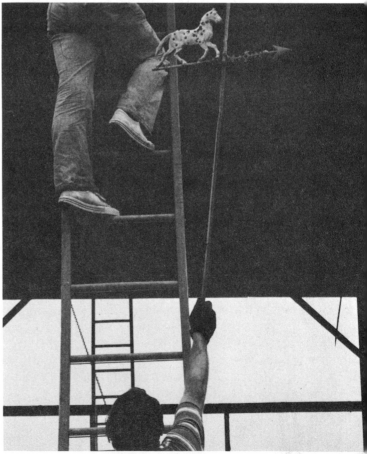

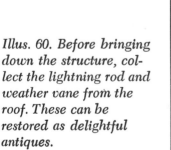

Illus. 60. Before bringing down the structure, collect the lightning rod and weather vane from the roof. These can be restored as delightful antiques.

Interior Salvage

The next step is to remove as much built-in material from the inside as you can. By built-in material, I mean the customizing features that lend a uniqueness to every barn. Most will have some kind of stalls for young stock or horses, breeding pens, rabbit hutches, or storage rooms that are entirely separate from the structure of the barn. The kind and quality of material can vary greatly, to anywhere from sturdy boards to warped plywood, or even paper-based boards. To remove them, you will find that the maul works best, and that most of these structures will fall apart with a couple of well-placed swings. Finish up with a pry bar.

Illus. 61. This barn has been partly floored over for a second level. You may need to remove stalls and interior partitions before starting on the main structure.

If the barn was designed for dairy cattle, look for varieties of stanchions used to hold the cows in one place. The earliest stanchions can be identified by their wooden rails, whereas more modern ones are all metal. The very early wooden ones add a rustic decor when they are placed around fireplaces or in family rooms.

Stanchions are held in place at top and bottom by short lengths of chain. The top chain normally attaches to a ring bolt, that is in turn fastened to the wood top rails. If you can unscrew the nut, the top is easily freed. The bottom of the stanchion is fastened by a chain to a ring bolt embedded in the concrete floor. Either cut the chain or, for wooden stanchions, unscrew the bolts that fasten the metal chain bracket to the bottom of the stanchion. Save the parts because you might want to reattach the chain bracket to the stanchion.

If you find small rustic gates or doors that have been exposed to the weather, pry off the hinges and save these intact. They are ideal for decoration. Remove any windows as described earlier under "Windows" (*see* page 48). Foundations often have small, decorative windows, too, so don't overlook these. Only when you have taken out every possible treasure from inside are you ready to start tearing down the building itself.

Illus. 62. Barn-wood gates can often be used just as they are for fences in your own yard. Some of the smaller doors on upper-level openings may be in good condition and will blend in perfectly with an elegant interior design for a modern home.

Taking Apart the Structure

Before you begin the real work, decide which of the two approaches you will use. The most common method used for barns is to remove the sheeting boards and some of the bracing, and collapse the remaining structure. The reason it is commonly done this way is that barn roofs are usually high enough to make most people uncomfortable, and the pitches of the gambrel roofs most commonly found on barns are tricky to walk on. There is a safety factor present in this method, because you will not be exposed to the danger of falls, either over the edge or through a weak spot in the roof.

One disadvantage of this method is that large beams may shatter and rafters may be sheared off along the wall that strikes the ground first. If saving every foot of usable material is your goal, and you have no fear of heights, take down a barn the same way you would take down a house. Start with the roof, remove the sheeting, and disassemble the rest of the structure as you move down from the roof. After watching the process of tearing off the roof as it stands, and doing it the other way myself, I would recommend collapsing the structure. It is far faster, and unless you need a certain number of especially long beams, you will lose very little.

As an alternative to both, one worker has used very tall tubular scaffolding to remove the roof from the inside. Though still rather far from the ground, scaffolding does offer a greater sense of security than hanging onto a gambrel roof. Since old barn floors are often weak, additional support is needed for the rollers on the bottom of the scaffolding. A pair of 2 by 12's (5.1 cm × 30.5 cm) set lengthwise to the barn will allow the scaffolding to be rolled back and forth with no danger of plunging a wheel through the floor.

Once you have the scaffolding erected on the 2 by 12's (5.1 cm × 30.5 cm), start removing the roofing from the inside just below the highest ridgeboard. Unless the roof has many layers of old shingles from several reroofings, the roof boards can be pounded off by using a 10-lb (4.5-kg) maul, punching through with the top of the head. Pound off all the roof boards, moving the scaffolding and 2 by 12's (5.1 cm × 30.5 cm) as you move down the roof towards the eaves.

When you reach the eaves, you will have to remove the eave board. Usually, this is wedged tightly against the lowest roof beam and is difficult to get out. With the roof boards gone you can work on it from above, but don't count on saving much of the wood since you may have to break it during the removal process.

Most barns have a hayfork carrier, a metal rail attached to the main roof beam for its entire length. These are both difficult and dangerous to remove while the barn is standing. To reduce the danger, tie a stout rope to a rafter, pass it under the metal track, over a rafter on the other side, and secure it to the lower wall. With one of these under each end, work with a helper to lower it slowly when it is free from the beam.

With the ends secured, start removing the metal fastenings that hold it to the beam. Metal brackets or clips may be fastened to the beam so the track can ride on a series of clips, or the track may be nailed or bolted directly to the beam. Use a pry bar, or whatever the fastenings require, to pry them loose. As the track drops, the ropes will support it. When you finish, lower the track slowly with a helper so it doesn't fall. The entire unit can usually be sold to an area farmer for use in another barn.

One valuable treasure taken from an old barn is the outer sheeting. Regardless of the approach you use, you must remove the sheeting while the barn is standing. Keep in mind that a high percentage of the building's total strength is supplied by the sheeting. A wobbly barn may collapse by itself as the sheeting is removed, so keep a sharp eye out for any signs of shifting.

Because the sheeting is easily damaged, it should be removed with care. Most barns are constructed with two levels of sheeting. The lower boards should be removed first so they will not be marred if the boards from the top level should slip as you are prying on them. When removing the outer sheeting, use either a nail puller from the front or a wrecking bar from behind. Pry as close to the nails as possible to prevent damage to the boards. If the boards are in good condition and are full 1-in (2.5-cm) stock, try striking them with the head of a 10-lb (4.5-kg) maul from the back. If the boards start to split, use the nail puller. Use a freestanding ladder whenever possible. If you must lean a ladder against the building, place padding between it and the boards to prevent scratching the natural finish.

As each board is removed, pound out all nails with the hammer, and pull them out with the wrecking bar as carefully as possible. A small, padded block under the head of the bar will help prevent damage. When all nails have been removed, the boards should be stacked. Set them in position, rather than sliding them in the usual manner for stacking lumber. The care applied to the surfaces of the weathered boards will prevent damage, and pay dividends later when the boards are either used in decorating or sold. I have watched careless workers toss heavy tools on the weathered surfaces, greatly lowering their value for decoration.

After removing all the lower boards, the top sheeting may be taken. Since the greater height demands a tall, heavy ladder, the boards will be damaged less when you work from the inside of the barn with the wrecking bar or a maul. As each board is removed, hand it to someone waiting below, because the boards are brittle and easily broken. When the sheeting is removed, only a skeleton structure of heavy timbers remains. If there is any wind at all, the structure will start to sway.

From the skeletal structure, the true design and construction of the barn are apparent. Note how the main beams are fastened together. If the building is very old, it may have been put together with wooden pegs called "trunnels." Look carefully at the beams to see if they show the marks of a broadaxe. If you find such marks, these are hand-hewn beams, highly desirable for decorating. The builder would undoubtedly be surprised to find his handiwork in demand as decoration today, since his only thought was of utility. The broadaxe was the only tool available to turn a round tree trunk into a square barn beam. By necessity, early farmers were expert carpenters; their ability to hew a round log square has been taken for granted. Anyone today who attempts to duplicate the hewing process quickly understands why the workmanship of the past is admired so much.

Illus. 63. Stack barn boards with the greatest of care, never sliding the boards over each other. See that all nails are removed before piling them.

Illus. 64 (left). The end of this beam has been cut to fit into a notch on its connecting beam. Illus. 65 (right). Notice the way these roof supports are connected. They are all notched and fitted with wooden pegs. The pegs can be driven out with a small dowel.

When everything has been removed, hitch a tractor, a four-wheel-drive vehicle, or preferably, two vehicles, to the highest beam with heavy rope or chain. By placing a slight tension on the ropes, you can control the direction in which the building will fall. With the tractors in position, start cutting off the angle braces between the main beams with a bow saw or chain saw. If the building is structurally sound, this is usually safe, unless there is a wind.

Next, saw halfway through the vertical corner-beams on the side opposite the vehicles, right at the floor line. At this point, the tractors should be able to pull the structure down—after you have left the building, of course. If they cannot drop the building, continue sawing and pulling alternately until the building collapses. Be sure to enter and leave by the side opposite from the pulling force, and see that all interested bystanders stay well away from the structure. When the building finally collapses, it will go down with a mighty crash, blowing clouds of dust and debris in every direction. The sight is spectacular, and it signals an end to the history of the barn. Your work, however, is just beginning.

Illus. 66 (above). Two tractors are hitched to the upper beams of this barn. The corner posts must be cut partway through. As the tractors pull, the corner posts buckle outward. Illus. 67 (below). The roof lands in a cloud of dust, ready for you to take apart at ground level.

Regardless of whether you have decided to collapse the building or tear it apart while it stands, you will remove the shingles the same way. The only difference lies in the height at which you will be working.

Barns were commonly roofed with cedar shakes. As the shakes became too rotted and cracked to prevent leaks, they were often repaired with sheet metal. On one barn, I found dozens of old license plates that had been nailed over the shakes. Sooner or later, though, the roof ages past the point of repair. The usual procedure has been to apply asphalt shingles right over the shakes. The resulting combination may be difficult to remove. Sometimes it may be easiest to remove the shingles first, and then work on the shakes. It depends on how the shingles lock together and on their condition.

Barn roofs are huge, so if you can save the shingles you will end up with a large quantity even if you damage some. To save them, work from the top ridge down on a warm day. When the shingles are off, work on the shakes from the eaves upwards. A wide, flat pry bar or an old ice chisel will cause the nailheads to pull through, and the shakes to fly off. Save any wide ones in a separate stack for artists to draw on or for decorative uses.

If you did not collapse the structure, you will find some roof boards beneath the shingles to pry off or pound free from the inside. These are usually spaced at least ½ in (12.7 mm) apart, to save lumber. Barn roofing boards are usually of low quality, but they can be used for shed roofs, hen houses, or pigsties. A flooring puller may work on roofing boards, too. Often, though, barn roofs used unsawed poles as part or all of the rafters. They may be 5 in (12.7 cm) round, so a tool designed to fit a 2-in (5.1-cm) joist will be of no use. Using the tip of a pry bar, start at the roof ridge and move down to the eaves. The angles may become steep, but you can usually use the roofing boards below to stand on while you pry.

The rafters will come out next. Barns are built with more variation in timber sizes than houses because there are no interior walls that must be smooth. Sawn lumber and hand-hewn timber can be found side by side with a variance of several inches in thickness. People used whatever they could get for barns, but they tried to see that the sizes matched in their houses. Some of the rafters may have been notched, fitted, and nailed to a side beam so strongly you will need powerful swings from a maul to budge them. Most barns are also built with a series of collar beams that connect the rafters under the ridgeboard. They are normally placed on every third or fourth set of rafters; pry or pound these off before taking out the rafters they connect.

Below the rafters, you will be dealing with the main supporting structure. Unlike houses, with their simple angles and toenailed joints, barns rely more heavily on the strength of their joints to hold the structure together. The resulting joints are often difficult or impossible to separate.

Different joints can usually be traced to a national preference. Experts can even recognize the styles popular in each European country. Some joints are much stronger than others, but may have taken longer to build.

A glance at the corner posts will indicate the strength of the construction. You may find the ends notched by axe or saw, and fastened with nails, pegs, or by the notches themselves. Where these are notched and nailed, the nails may be ⅜ in (9.5 mm) thick and a foot (30 cm) long, driven in by a maul. I know of no ordinary equipment that can take such joints apart. The only solution is to leave the joints intact, sawing the timbers off on each side of the joint.

Sometimes the cross-bracing and the corner posts will be joined with the end of timber carved down to fit into it. If all the bracing has been designed this way, start on one end of the building. After you remove one of the corner posts, remove the other braces in turn so the plugs can be removed from their sockets. You can salvage all the supporting structure without breaking or sawing anything.

If you are working on the structure while it stands, lower the beams slowly to prevent damage. For the heavier beams of oak or elm you will need ropes to lower them. Paired ropes, looped over lower timber on each side, may be the best procedure. You can usually lower the ends one at a time as they are freed.

Where the structure has been built using trunnels, the joints are not only easily disassembled but also highly decorative. Use a hammer and a small piece of dowel or metal pipe to drive the pegs out the same way they went in. They drive easily due to their many years of drying and consequent shrinkage. Sometimes the pegs are not driven all the way through. In this case, grasp the protruding portion with a large pair of pliers and pull them out by pulling and twisting. Save any pegs you take out because they may be useful to join portions later, or to use for decoration. Where peg holes appear in beams used as supports or decoration in remodelled rooms, the old pegs look far better as hole fillers than new pegs.

Floors

The typical barn will consist of a ground-level or slightly above ground-level wood floor and a lower floor of poured concrete. The wood floors were built to withstand the weight of heavy tractors or tons of hay, so the floorboards are likely to be sturdy enough for the task. Two layers are common, with 1-in (2.5-cm) boards below and 2-in (5.1-cm) boards over the top. These are exceptionally heavy timber, useful for many types of new construction. One word of caution applies, though. If animals have been kept on the wood floor, the odor will never leave the boards; therefore, the boards should not be used as supporting or flooring lumber in a house. You will encounter this problem especially with horse barns. Flooring in horse barns may have little odor in dry weather. During a rainstorm or any humid day, however, the odor is strong enough to make you wish you had never used the boards as part of your house. It is wiser to use such boards for outbuildings, or to sell them with the understanding that they will be used for that purpose. The same thing applies to any wood used for horse stalls.

Removing the floorboards is seldom any problem. The flooring may have been nailed sparsely, due partly to the rigid double layer that flexes so little it can hardly shift, and partly due to the cost of nails. Usually, an occasional pry with the pry bar will quickly free the boards. Carry them outside and place them in a huge stack.

Joists

In many barns, the floor joists are 13-in (33-cm) round poles. The only practical way to deal with them is to arrange for help. Even softwood joists are terribly heavy when they are 25 to 30 ft (7.6 to 9.1 m) long. Lacking help, you would need either a winch or a chain saw to cut them into manageable length. Cutting them up is a waste of good lumber. Even if you decide against using them full-size, they can be hauled to a sawmill and cut to more common dimensions.

The spacing of floor joists is often irregular. Where the builder planned for tractors and other heavy machinery to be driven, the joists will be spaced very close together. Outside this area, the spacing between joists will sometimes more than double. This design allowed the builder to maintain strength where it was needed, yet saved material on the other sections.

Beams

Beneath the joists, you will find huge supporting beams. These pose even more of a problem than joists. If they are hardwood, they weigh more than even three workers can handle. Use a winch or cut them into shorter lengths. Before cutting a beam, check the area closely for nails; resharpen your chain if it strikes one. Consider yourself more fortunate if the beams are of softwood that is light enough to enable two men to carry one to a trailer and load it. Be sure to read the section on hauling salvage before transporting unusually large beams.

Posts and Columns

Beneath the main-floor beams, some type of support columns will be left. Most likely, these are of wood, but metal supports will occasionally be found. Both are useful, since the wood supports can be used for gate or fence posts in rustic yards, and the metal columns can be used again as supports or sold for use as scrap iron. With these removed, only the foundation remains to be knocked down if it protrudes above ground. Then the hole can be filled in for the final landscaping.

Final Jobs on the Salvage Site

When to Haul

At some point you will be hauling away the lumber and other materials you have salvaged. Ideally, of course, you should own your own truck or van. Then you can stop tearing down early enough to spend some time sorting and hauling the day's salvage; make sure nothing is left behind for thieves. Most workers follow this procedure, but under some conditions the salvage may be left until later.

One reason for leaving salvage behind is that the building being torn down is nearly inaccessible for vehicles. This might be the case with cabins or farms in wilderness areas. People rarely steal much lumber if they have to carry it any distance, so if the access road is poor, you might prefer to leave the lumber behind until you are finished tearing down. Then you can repair the road as needed, and drive in and load everything over a short period of time. This approach is especially wise if you plan to rent a truck later for the hauling process.

It would also be best if the storage site you plan for the salvaged material is not at your own home. This way, you won't be wasting trips to and from your home when you could be hauling materials on each return trip.

Techniques of Hauling

Unless your truck is unusually large, you need every bit of space to haul away the salvage for one day. Start loading the longest materials first so everything else will rest on a firm base. Be sure to wear gloves whenever you handle lumber. The nails sticking out everywhere prevent the boards or timbers from fitting close together, so each board will occupy more than twice the space it should. Under some conditions, perhaps where the hauling distance is great or your vehicle is small, pull out the nails before loading, to save space.

After you have piled in the longer materials, continue piling in the shorter boards until you have a full load. A full load does not mean that the load juts over the top. If you must go over hills, the upper materials tend to fall off, presenting a hazard to vehicles behind you. As a precaution, lash an old piece of canvas over the smaller boards to be sure nothing falls off. You don't want fallen materials to cause an accident. If the hills are steep, tie in some of the larger lumber on the bottom just to be sure the whole load can't slide out. Even with these precautions, it's a good idea to stop occasionally on a long haul just to see that the load is riding well.

Illus. 68. Typical salvage from a cabin. These items can all be used again.

Illus. 69. Tongue-and-groove pine boards taken from the interior of a cabin are welcome additions to a modern home.

Among the longer materials you may have to haul are long sections of piping from large buildings or barns. These may be more easily used if they are in one piece, so try to keep them intact. Rather than setting them in the box of a pickup truck, and allowing parts to stick out of the box, try loading them on top of the cab. Use some foam padding to prevent damage to the top, and tie the ends to the front bumper and the tailgate. Then the pipes will not stick out so far in back. As you drive with any unusual loads, keep an eye on the rearview mirrors to check for anything that might be jarred loose.

Whether you store salvaged materials at home or at the demolition site, be sure they are protected from rain. Pile them up and cover them with a tarp weighted along the edges by old bricks or stones; this way the wind can't blow them away. Old wood quickly rots if it becomes wet, so do everything you can to keep it dry.

Illus. 70. Tie all lumber on securely before transporting it. Make sure your load complies with the road regulations in your area.

Nail Removal

When you have transported all the salvage to a storage place, pull out the nails so they can be used again. Removing nails is time-consuming, but relaxing. There is hardly any chance of getting hurt, so you don't have to think every minute about what you are doing—as you must when tearing down a building. Use a medium-sized claw hammer and a wrecking bar with a nail puller. Lay the boards between two sawhorses so they have support while the points are hammered straight enough to pound back through. Then turn the boards over and pull them out by the heads all at once.

Some workers carefully save and straighten every nail. I met one worker who built an entire garage without buying a single nail. I have met others who consider it a waste of time because all their construction is done with power nailers. You will have to decide which idea makes sense to you. Check the current prices on nails, estimate how many you will need for your construction project, and you will be able to determine if the amount you will save is worth your time.

Storage

After removing all the nails, you are ready to sort and store the salvage. Depending on the size of the building, you may have different quantities to store. With a small amount, sort just by size, keeping all 2 by 4's (5.1 cm × 10.2 cm), all 1 by 8's (2.5 cm × 20.3 cm), etc., together. With large amounts of salvage, you might also sort by quality. Boards with large knotholes or splits should be stacked separately from top-quality boards to make selling or reusing them simpler.

However you sort them, be sure you allow for enough air circulation between the boards. Keep the bottom layer of boards off the ground or floor and place ½- to 1-in (12.7- to 25.4-mm) scrap sticks between each layer as you stack. Otherwise, dampness may cause rot, especially if the wood is damp from a rainstorm. If you have no indoor storage area, be sure the wood is all covered with a waterproof tarp and stacked high enough above the ground line so runoff water can't reach it. You have worked hard to tear the building down, so take care to protect the salvage.

Many times, workers are exhausted when the final hauling is done and the last load is empty. They plan to cover everything tomorrow, but the next day brings more problems or work, and in no time at all rainstorms have soaked everything and the wood is covered with mould. I have seen incredible wastes of material due to improper storage by people who never followed the last step of proper storage. Make sure it doesn't happen to you.

Selling Salvage

Depending on your reasons for tearing down the building, you may or may not plan to sell the material. Most workers sell or trade at least part of the salvage, and here again, you will have to decide what works best for you. Any material to be sold is best sold at the site, rather than hauling it somewhere else to sell later. I have talked to other workers, though, who prefer not to be bothered by buyers while they are working. If you are working high up on a tall building, it would be a problem to keep climbing down to deal with a flood of customers. So if you plan to sell materials at the site, it is better to wait until you have the upper parts all off, and are working on the lower walls or floors.

If you advertise the fact that you have salvage for sale by erecting a sign or giving an address in the paper, you are inviting theft as well as potential buyers. Some workers feel that if few people know they are tearing down a building they will be spared the danger of theft. It seems to depend on the kind of building and the location. Obviously, some locations lend themselves better to theft than others. Only experience by trial and error can tell you what to expect in your area.

Sometimes you can sell a particular kind of salvage to the right buyers.

If you have a quantity of shorter lengths of scrap wood, you could run an ad in the paper such as:

Attention Farmers:
Scrap wood for pigsties, hen houses, etc.

If you place only a phone number, you will not encourage people with dishonest motives. Try the same approach with plumbers, handymen, or heating specialists, depending on what you have to sell.

If you want salvage items to sell well, keep the price low. Twenty to thirty percent of the price of new material is common, giving you that much profit and the buyers a 70 to 80 percent saving. Windows and doors usually sell well, as the prices on these items are high when purchased new. Check on the current prices of everything, and price your salvage accordingly. Some buildings may have been built with unusually heavy timber that are nearly impossible to find today. Check with sawmills to see what they would charge to cut beams like them. For special beams, the percentage you charge should be much higher, closer to 60 percent in some cases.

You will be well paid for all your work in removing and hauling the timber. Remember, though, that the ease with which salvage can be sold is dependent on the economy. When no one is building, everything is harder to sell. Due to the lower cost, though, salvage may still be selling when the sale of new materials has all but stopped. You will find people in desperate need of a garage or storage shed who have been holding off on building because of the high cost. Salvaged materials may provide the answer to their problem, and the solution to yours.

Cleanup

With the last useful materials removed, complete the tearing down of the building. If the contract you signed specifies that you remove everything down to the foundation, you are finished with the project. If you are responsible for the foundation, you will have to deal with that next.

Some foundations are so flimsy you can knock down whole sections with a few swings from a maul. If the foundation is in good condition, though, you will need some power machinery. A tractor with a backhoe works well, and so does a bulldozer, if you can get one for a reasonable price. Usually, you will be able to push all parts of the foundation into the basement, thus reducing the foundation to ground level. If the building does not have a basement, truck the foundation away just as you have done with the other debris.

The final cleanup can be a tedious job. If you intend to burn the rubble, plan accordingly. In most localities a burning permit is needed. Check with the fire warden or official who handles fires. Be sure you don't try any burning unless the winds are almost dead calm, because the sparks can carry long distances.

Illus. 71. Only a few days ago this area was covered by a large building. Fill was hauled in to cover every trace of the original foundation, and levelled out with a tractor.

If you plan to burn old sections of hardwood beam, you will be disappointed at the length of time needed for them to be consumed. Softwood burns much faster, often with a sudden flash of heat and great showers of sparks. Be sure there are no valuable shrubs or trees for at least 25 ft (7.6 m) in all directions, because the heat can either kill them or cause them to catch fire. Even at best, the burning of rubble may take several days of constant watching. At worst, I have found that hauling debris from hardwood beams to a dump up to 18 mi (29 km) away was easier and faster than burning. Try both techniques to see which you prefer.

One unpleasant surprise to most workers is the vast bulk taken up by old cedar shakes. Normally, these are seen either in neat bundles or applied to a roof. But when they are suddenly thrown into a mound during the demolition process, they form whole truckloads that refuse to be condensed.

To gather up old cedar shakes or even fragments of asphalt shingles, build a carrying frame by stretching canvas between two 3-ft (.9-m) poles. Pile debris on the canvas and use the poles as handles. The technique works fastest with two or more workers. Sometimes an old door—and two people to carry it—can work well for hauling small broken boards or other debris. No matter what you use, though, the final cleanup will take longer than expected.

If the basement becomes filled, haul the rest to a dump. Years ago, small township dumps were common. Today, however, dumps have become more consolidated in some states, and an entire county may maintain only a single dumping site. If the dump is on the opposite side of the county, you will have a long round-trip haul. It might be worthwhile to rent a truck larger than the van or pickup you own. Depending on the vehicle you already own, the amount of rubble to be hauled, and the distance involved, the rental may or may not be practical.

If you have agreed to complete the landscaping, haul in fill sand, which requires some kind of power equipment to spread and smooth. The process takes only a short time for a skilled operator, so unless you own the kind of equipment that is needed it is best to hire someone to do it. Many people own tractors or four-wheel-drive vehicles with blades that can be used to landscape an area. If you already own one of these, you can do the job yourself, and complete the process.

With the cleanup completed, you are ready to think seriously of using the salvage as building or decorating materials in your home.

Using Salvage in
Room Design

After you have successfully dismantled your structure, the interesting work of incorporating your salvaged materials into your home begins. As you worked on the salvage project, you probably formed many ideas of where you would like to use certain items in your new building. The variety of practical and whimsical uses for salvaged materials seems boundless. Your personality and preferences will determine how you use the salvaged materials in your home.

On the following pages, we show how some people have used beams, boards, bricks, and other items salvaged from old buildings to enhance the beauty of their homes. Hopefully, these examples will inspire you to work salvaged materials into your home using both designs unique to your building and the salvaged materials you have collected.

Salvaged Boards

Weathered wood is a unique decorating material. Because it is not readily available in lumber yards and building-supply stores, the decorator who uses weathered wood is assured that his interior designs will have an unusual appeal. Naturally weathered wood is usually deeply grooved. Colors vary from chocolate brown to charcoal grey. Some boards retain some of the original color of the paint once used on them. Many boards salvaged from barns, for example, are streaked with red.

Weathered wood is most popular used as picture frames or panelling, but it can also be used any other place you would use boards for building or decorating. Even its use as panelling varies greatly. You can panel an entire wall in vertical boards or use the boards halfway up the wall for a wainscoting effect. You can arrange the boards vertically or diagonally along the wall. Brick or chevron inserts lend variety to a long wall panelled in weathered wood. Illustrations 72 to 82 show salvaged boards used as panelling, fences, doors, and stairways.

Illus. 72. Salvage can be used in a variety of ways. Here salvaged beams are used for trim around the second level of this open room. Salvaged boards panel the upper part of the walls. This is a variation on wainscoting, which usually starts at the floors and extends halfway up the wall, rather than beginning at the ceiling as this panelling does and extending downwards. A folding door made from barn boards separates the dining room from the open living area. A light salvaged from an old church completes the decor.

Illus. 73. A close-up of the folding door made of barn wood seen in Illus. 72. Notice how the boards are hinged together.

Illus. 74. These salvaged barn boards are used as vertical wall panelling and ceiling panelling.

Illus. 75. This is an example of panelling placed diagonally on a wall. Notice how well this panelling fits into the corner of the room. The corner is finished off with trim moulding cut from another salvaged board.

Illus. 76. Here barn boards are used for wainscoting, a variation of panelling. When using salvaged boards in this way, keep in mind the material used above the wainscoting. In this case, the rich brown tones of the boards are highlighted in the colors of the wall-paper used.

Illus. 77. Barn wood can be blended with logs to create unusual room design. Here the logs are used on the bottom half of the wall, and the boards for the top section. A brick insert, visible on the left, adds variety to the room design.

Illus. 78. Here barn boards are used for panelling and a storage box. The old hook and hinges salvaged from a barnyard gate are examples of the decorative touches you can add with miscellaneous salvage. During any salvage process, keep your eyes open for small items such as hinges, hooks and square nails which can become some of the true decorating treasures.

Illus. 79. This cupboard is constructed from barn boards. In keeping with the weathered wood, leather is used for hinges and a wooden clasp keeps the door shut.

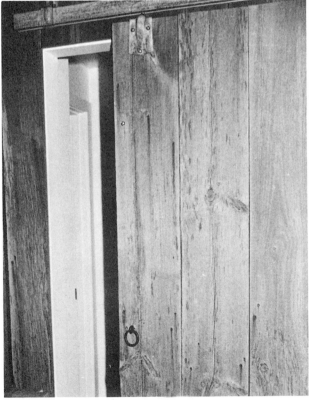

Illus. 80. This door—built from salvaged barn lumber—is used as a door to an upstairs room. One unique aspect is that the door slides, rather than swings, open. A new slide and fittings were purchased at a local hardware store, and the metal parts were aged with hydrochloric acid to complement the weathered wood. Because this door opens off an upstairs balcony, it looks very much like an authentic barn door.

Illus. 81. Even the stairway in this house is constructed with salvaged boards. A handrail, salvaged from a stairway leading to a hayloft, completes the stairs.

Illus. 82. Here barn wood combines effectively with brick to make an attractive fence and gate.

Salvaged Beams

Beams are another popular salvaged material. Beams can be found in barns, schools, warehouses, and other sturdily built older structures. In new structures, they are sometimes used as ceiling rafters, support pillars,

118

or for a number of other decorating purposes, but their most common use is for fireplace mantels.

Beams are special for two reasons: the type of wood used for the beams and the size of the beams. Most barn beams, for example, are from 6 to 13 in (15 to 33 cm) square, a size of lumber almost impossible to find in sawmills today. The beams are also often hardwood, which makes them even more rare for today's builder. Because of the scarcity of such building materials, they are invaluable for many building purposes.

Another special feature of many beams is their aesthetic value. Many beams were hand-hewn. With the axe as his only tool, the pioneer builder shaped a giant tree into a square beam. The axe marks are still visible on these beams, and today people marvel at the precision of the axeman who was able to shape the boards as square as any modern headsaw. If you find a hand-hewn beam, you are extremely lucky.

Illustrations 83 to 90 show salvaged beams used as fireplace mantels, ceiling rafters, and even a hearth. Their uses are bounded only by the imagination of the decorator who works them into a room's design.

Illus. 83. Beams can be used together in a variety of ways. Here elm and yellow birch beams are used for a ceiling cross-bracing. Another beam, which serves as a mantel, was set into the brick as the fireplace was built. In this home, beams are also used for the fireplace hearth; notice how they have been arranged so the grains match in the front.

Illus. 84. This fireplace mantel was constructed from a barn beam and miscellaneous salvage from different barns. The heavy beam is suspended by lightning-rod cables. Sturdy hooks, common in barns to suspend the pulleys for hoisting hay bales or other heavy equipment, were fastened into the beam. Then the cable was fastened to the hooks. The cable has weathered a dull black so it matches the wrought-iron trim on many fireplaces. To hang the beam, holes were drilled in the ceiling. The lightning-rod cables were passed through the holes and wrapped securely around collar beams in the attic.

Illus. 85. *This beam rests on a brick shelf. It is bolted to the wall; then new wooden pegs are used to cover the openings drilled for the pegs. The pegs blend in very well with the rest of the beam, almost as though they were salvaged as part of the beam itself. Notice the weathered wood used as a background for the metal sculpture hanging above the mantel.*

Illus. 86. *Here is a way of supporting a beam used as a fireplace mantel. Two bricks on each side of the mantel are set at right angles to the fireplace wall. A bolt, passing through the natural hole in the brick, is used to hold the beams firmly in place.*

Illus. 87. The beams in this room, visible from an upper balcony, provide solid yet rhythmic effect. Every angle seems to be repeated continuously.

Illus. 88. Where beams are not long enough to reach across a span, they can be spliced together beneath a support post. This splice has been fitted in the traditional manner with a zigzag notch cut from each beam. You have to look carefully to notice the splice; it blends naturally with the original axe marks on the old beam.

122

Illus. 89 (above). Hemlock beams salvaged from a school dating from about 1895 were used to construct this contemporary home. They have been extended through the end wall to provide a protective overhang for a deck below; the center beam fits through a notch cut from the support beam below it. These old beams blend in very well with the new lumber used for the ceiling. Illus. 90 (below). A close-up shows how these beams form a solid "T" over the patio doors.

Salvaged Brick

Like the boards of older buildings, salvaged brick is valued for its weathered appearance. The variety of shades, which result from time and exposure to the elements, lend a warm glow to a brick wall or fireplace that cannot quite be duplicated with new brick. Old brick can be quite fragile, however. Illustrations 91 to 93 show some hints on cleaning the brick, and some suggestions for using them in your room design.

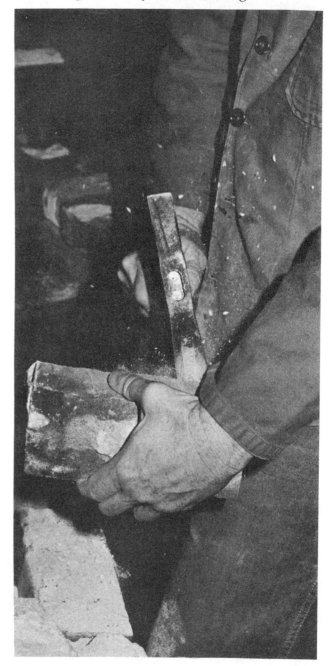

Illus. 91. Old bricks are expensive, so you will want to save as many as possible. They are also fragile, so you will have to work carefully. A brick hammer is the best tool for removing old mortar. If you attempt to clean the bricks on a hard surface, however, you will find that they chip or break easily. Hold each brick firmly in your hand, as shown. Then strike the mortar from the brick with the brick hammer.

Illus. 92. These bricks were salvaged from an old school. Half bricks are used above the mantel for heat vents. These bricks were cleaned differently from the ones in Illus. 91. They were rubbed back and forth on a cement block with sand on it to remove the mortar from the brick. The mantel, a salvaged beam from the same school, is set into the wall the width of the bricks.

Illus. 93. These bricks were salvaged from an old pottery kiln. Notice the variety of the shades of the brick.

Miscellaneous Salvage

Every salvage project turns up miscellaneous items such as lightning rods or milking stools that are eye-catching. You can't resist lugging them home, but once you have them you wonder what to do with them. Many of these items can be used in your room design as conversation pieces. Although they may not have the functional uses of boards, beams, or bricks, they are important to your decorating scheme. They lend variety and interest to room design. Illustrations 94 to 96 show a few of these items used as room decorations, as well as a reconstructed cabin.

Illus. 94. This barn ladder has been successfully incorporated into contemporary room design. As it runs along one corner of the room and disappears above the rafters, it looks very much as it did in the barn from which it was salvaged. You almost feel as though the soft, sweet-smelling hay awaits you just above the last beam. Note the axe marks on the hand-hewn beams and the wooden pegs used to hold the beams together in the original barn.

Illus. 95 (above). A wooden cow stanchion, worked into the decor of this house, has been inserted into the fireplace hearth. Here it provides a railing for the hearth and also serves as a conversation piece. Illus. 96 (below). Sometimes an entire building is salvaged and reconstructed in a different location. This cabin was built from white oak logs in 1860. Many years later, it was dismantled log by log and moved to its present location, where it is used as an artist's studio. To make sure that the logs were fitted into their original places, each log was marked with a metal tag bearing the correct number. Then the building was reconstructed following this numbering system.

Index